War and International Relations

This book provides a critical overview of the occurrence of war in the international system by examining the concept from multiple perspectives and theoretical backgrounds.

War is an essential concept in international affairs, if for no other reason than because prevention of war requires an in-depth understanding of it as a concept. This book seeks to examine the continued occurrence of war in international relations, despite the emergence of arguments concerning its obsolescence. It provides a new cognitive framework through which to understand war as a phenomenon, which can be applied to real-world scenarios and policy issues, making use of case studies predominantly from China and Japan. Theoretically, the book is primarily based on a structural realist framework but adopts a significant constructivist component through the emphasis on identity and reputation in the international system. The volume offers a nuanced yet holistic approach to the theory of war and seeks to engage critically with the major theoretical approaches, pointing out the major criticisms of these ideas and how the theories correlate.

This book will be of much interest to students of strategic studies, foreign policy, and International Relations.

Balazs Szanto is a lecturer at Chulalongkorn University, Thailand.

Cass Military Studies

Military Strategy of Small States
Responding to External Shocks of the 21st Century
Håkan Edström, Dennis Gyllensporre, and Jacob Westberg

Understanding Insurgent Resilience
Organizational Structures and the Implications for Counterinsurgency
Andrew D. Henshaw

Military Strategy of Middle Powers
Competing for Security, Influence and Status in the 21st Century
Håkan Edström and Jacob Westberg

Military Mission Formations and Hybrid Wars
New Sociological Perspectives
Edited by Thomas Vladimir Brønd, Uzi Ben-Shalom, and Eyal Ben-Ari

French Defence Policy since the End of the Cold War
Alice Pannier and Olivier Schmitt

Russia's Armed Forces
A Modern Military for 21st Century Warfare?
Rod Thornton

For more information about this series, please visit: www.routledge.com/Cass-Military-Studies/book-series/CMS

War and International Relations

A Critical Analysis

Balazs Szanto

Routledge
Taylor & Francis Group
LONDON AND NEW YORK

First published 2022
by Routledge
2 Park Square, Milton Park, Abingdon, Oxon OX14 4RN

and by Routledge
605 Third Avenue, New York, NY 10158

Routledge is an imprint of the Taylor & Francis Group, an Informa business

British Library Cataloguing-in-Publication Data
A catalogue record for this book is available from the British Library

Library of Congress Cataloging-in-Publication Data
Names: Szanto, Balazs, author.
Title: War and international relations : a critical analysis / Balazs Szanto.
Description: London; New York, NY : Routledge/Taylor & Francis Group, 2021. |
Series: CASS military studies |
Includes bibliographical references and index.
Identifiers: LCCN 2021001237 (print) | LCCN 2021001238 (ebook) |
ISBN 9780367748005 (hardback) | ISBN 9781003159605 (ebook)
Subjects: LCSH: War. | International relations. | Military policy.
Classification: LCC U21.2.S93 2021 (print) |
LCC U21.2 (ebook) | DDC 355.0201–dc23
LC record available at https://lccn.loc.gov/2021001237
LC ebook record available at https://lccn.loc.gov/2021001238

ISBN: 978-0-367-74800-5 (hbk)
ISBN: 978-0-367-74802-9 (pbk)
ISBN: 978-1-003-15960-5 (ebk)

Typeset in Times New Roman
by Newgen Publishing UK

Contents

Illustrations

Figures

Maps

Tables

1 Introduction

This book seeks to provide a compilation of theories, practical concerns, and methodological tools to critically analyse war in a foreign policy context, both from explanatory and normative perspectives. It contains all the nuts and bolts that have been deemed important to construct an analytical model with sufficient validity to be useful for the study of both foreign policy and foreign policy formulation. If nothing else, the book is a concise reference guide to the various theories and approaches to the study of war, offering a launching point for an in-depth study of the subject.

War: The reports of its death are greatly exaggerated

One of the key points of this book is that war continues to be a relevant part of foreign affairs because of the unique nature of war as a foreign policy tool for which, to date, we have found no replacement. Most foreign policy options are based on bargaining or overcoming the opponent's will in some fashion. Even coercion relies on the ability to overcome the opponent's will to resist. The use of force is the only major tool in a state's arsenal that allows it to directly seize and hold objectives through a test of strength, bypassing the opponent's will as a factor. Subduing the enemy's will in a war is an important factor in conflict termination but, theoretically, if one possesses sufficient strength, one could hold an objective indefinitely even if the opponent refuses to admit defeat. This is one of the key reasons that war continues to be present in the international system: We simply don't have a substitute for it.

There is a school of thought in academic International Relations that argues that war is dead. These arguments focus on a combination of nuclear armaments, the high costs of war in general, reflections on the two world wars in particular, and the evolution of international norms. Mueller argues that, just as humanity has abandoned slavery or duelling, society has left war behind.[1] There is a general view that the world is becoming safer and more peaceful as civilization naturally progresses forward, an argument supported by Pinker.[2] This thesis is obviously appealing: It appeals to one's natural belief that humanity constantly improves, every generation reaching closer to an ideal utopia than the one before. It is a very common sense understanding of history: Learning from mistakes and continuously improving. In opposition to this, there is empirical evidence that the system as a whole is not becoming more peaceful.[3] War and use of force continue to be present in international relations. In 2014, Russia annexed the Crimean Peninsula and engaged in a proxy war with Ukraine. The Syrian civil war has been raging since 2011, seeing intervention from Russia, the United States and Turkey. In 2020, China and India engaged in border skirmishes in the Himalayan region and Azerbaijan and Armenia fought over disputed border regions. This idea is often waved away by the argument that war has only diminished in the developed world, while the developing world still awaits enlightenment.[4] This alone would provide sufficient reason to continue the study of war as an intellectual discipline, since a large part of the world's population resides in developing countries. But arguably war has not been banished from the developed world either. The United States has waged two long and bloody wars in Iraq and Afghanistan. Developed countries went to war in Libya. It is true that we have not seen a major-power war comparable to the two world wars. But that would be a narrow interpretation of war. The idea of subduing an uncooperative opponent through might alone remains appealing, especially when developed countries often possess disproportionate military power relative to their would-be opponents. The 2003 war in Iraq was short (the initial combat phase) and it witnessed the overwhelming force of modern joint operations warfare. But it was a war, nevertheless, when examined from a foreign policy perspective.

Even if one would seek to restrict the argument to major power wars, one should not mistake temporary bubbles of peace for the demise of war. The bipolar dynamic of the Cold War froze tensions, but they were channelled through proxy wars nevertheless. And at the end of the Cold War, there was a sudden onset of a unipolar order, accompanied by the economic exhaustion and political collapse of the major antagonist. There was an absence of major power confrontation as the ideological victory of Western liberal market economics seemed absolute. Yet today one can observe the rise of China as Beijing's attempts to establish its own sphere of influence. While traditionally China has acted within the limits of the existing system, its revisionist intentions, growing dissatisfaction and insecurity is increasingly evident. Once deemed unstoppable, liberal values are on the back foot in many countries, heralding in a new era of competition and political polarization in the international system. Under such conditions, it would be premature to declare that war is now unthinkable amongst major powers. The conditions that permit the occurrence of war are still present, waiting for the optimal conditions to once again unleash the tragedy of war onto the system. The nature of international relations has not changed so fundamentally that war as a tool would have lost its utility. It is not a universal tool. It is applied to niche foreign policy problems. But those problems still potentially exist.

Thus, this book argues that the reports of war's death have been greatly exaggerated. In the following parts, this book presents its argument as to why wars still can occur within the system and how rational policy calculations could lead a country to make the conscious choice to go to war. Ultimately, war is never "defeated". That would be akin to arguing that fire or the weather is defeated. War can only be carefully managed and minimized through strong preventative policies. However, this requires a continued critical study of the subject, which is the purpose of this book.

The importance of theory and modelling

An unfortunate fact for foreign policy formulation is that reality is infinitely complex. Events and outcomes in the international system can be hard to explain, let alone predict. While Chaos Theory would argue that regardless of the seeming randomness of the process, there is an underlying cause-and-effect pattern, such an avenue is of limited utility for policy analysis. Rather than to conduct the analysis on this infinitely complex reality, this book focuses on constructing a sufficiently complex model to substitute for it. Constructing a model allows one to highlight high-impact factors while filtering out low-impact ones that would clutter up the analysis. This allows one to construct an analytical framework that balances validity with pragmatic utility.

The main trade-off for constructing models is reduced validity especially in an anticipatory or normative predictive context. Simplification will always reduce the accuracy of the results gained from the model. One can imagine a spectrum, where on one end there is reality in its infinity complexity and on the other end there is a highly simplified model. Moving from one end to another accuracy will decrease, eventually reaching a point where it is no more accurate than simply guessing. At the same time, moving in the same direction ease of use increases. The goal is finding the 'Goldilocks zone' where the results are accurate enough to be useful, while complexity is kept to a reasonable level.

This book primarily focuses on the construction of a rational-choice analytical model, based on a bounded rationality framework. The goal is not only to create a practical model for decision-making, but to create a model of the international system in which that decision takes place. In the later chapters, this book seeks to walk the readers through all the key steps of and considerations for the construction of this model. The objective is to create a framework where data about a particular decision-making situation can be input on one end and a reasonably accurate answer can be outputted on the other end, allowing for both explanatory and normative/anticipatory conclusions to be drawn. In accordance, this book is intended to be practical, sometimes at the expense of more abstract grand theoretical debates in International Relations.

That is not to say that theory is unimportant. International Relations theory provides a crucial foundation for the construction of any model. Theory can guide one to determine which factors are potentially high-impact ones that are necessary to include in the model and which ones can be potentially disregarded. Theory is also essential in providing a framework to how to analyse these factors, what data needs to be prepared for the model to work, and how various components interplay. This book builds on existing International Relations theory to construct a compelling analytical model for the decision to go to war. To keep the theoretical engagement simple, the book looks at specific thinkers and theoretical works, rather than at larger theoretical traditions as a whole. This is a deliberate attempt to make the book more accessible for those with limited to no background in International Relations theory. The book is not intended to be the end of all arguments on International Relations theoretical debates, rather a launching point for a critical analysis on the continued occurrence of war. The readers are highly encouraged to seek out deeper engagement with any of these schools of thought if their ideas have piqued an interest. Unfortunately, such endeavours are decidedly beyond the scope of this book. Rather, this book is envisioned more as a guide: A collection of ideas, concepts, and methodological tools students of war and foreign policy can come back to for quick reference. One of the purposes of this book is based on the idea of how to allow students to successfully utilize International Relations knowledge, not necessarily in an academic but a practitioner's context.

The importance of foreign policy and preventative diplomacy

This book is not an endorsement of war. Just as Mearsheimer's recognition of the inevitability of the tragedy of great power competition has been mischaracterized as warmongering, this book's – and the courses that it is based on – critical discussion on war has been sometimes misinterpreted as an advocacy for war. A popular saying, often attributed to Pericles the Greek statesman, is that even if you have no interest in politics, politics might take an interest in you. The same applies to war: Even if one has no interest in war, war, even if only with its absence, will nevertheless affect their life. Foreign policy, including the decision to go to war or not, determines the fate and fortune of nations, shaping the lives of all of its inhabitants in the process.

Even if one considers war to be an unacceptable policy outcome, a blanket rejection of the field of study, its banishment from international relations study, or the declaration of a moral victory will not serve the creation of a more peaceful international system. Wars have been and are happening. Just during the preparation of this chapter, Armenia and Azerbaijan engaged in a military clash over the sovereignty of the Nagorno-Karabakh region. One's best hope to create a relatively peaceful international system is to engage in effective preventative diplomacy. It will not prevent all wars – some wars are a result of a combination of insurmountable structural forces – but it will minimize their occurrence. However, to effectively engage in preventative diplomacy, one needs to understand the conditions and reasons that lead to a war in depth. One can view policy formulation as a chain of rational calculations. Preventative diplomacy seeks to alter these calculations by intervening in key points, raising costs here, lowering benefits there, balancing offering concessions and taking a deterrent stance, all in order to produce a preferred policy outcome of not going to war. Under such conditions one needs to not only understand the politics of one's own state, but that of the opponent: The policy manipulations largely target the will of the opponent so they end up making a different choice.

This book seeks to assist the reader in creating the critical theoretical and methodological framework to support effective preventative diplomacy. The main points are (a) to critically examine the theoretical contexts of war to fully understand why wars keep happening and why one cannot dismiss the phenomenon as a relic of a less enlightened age and (b) to demonstrate how a rational choice model can not only be used to analyse why a country chooses to go to war but as a methodological foundation for effective preventative diplomacy by identifying key pressure points within these calculations.

Structure of the book

The book is divided into four main parts. These parts seek to address issues related to the occurrence of war in the international system from both a theoretical and a pragmatic, policy-oriented perspective. In the theoretical parts, the book looks at specific selected works that are representative of the larger school of thought, rather than trying to engage with the school of thought as a whole. This is done to ensure that key tenets can be identified clearly as there are often minor variations from thinker to thinker that can distract from the overall issue at hand. Thus, key works have been selected to present a single set of rules that can be subjected to critical engagement. If the reader is interested in a specific school of thought more broadly, then this book can be a good starting point, but further research will likely be needed.

This book's analysis on the question of going to war begins with examining war itself. Part I focuses on examining a diverse set of questions and contexts in which war can be understood. The fundamental thesis of this part is that war should be regarded simply as a tool of statecraft, a part of a larger foreign policy arsenal, rather than something wholly distinct from politics and diplomacy. Part I is divided into three chapters. Chapter 2 seeks to engage with the definition of war to demonstrate the value of the interpretation of war as a foreign policy tool rather than something distinct from foreign policy. Chapter 3 examines the ideational and cultural context of war. The main focus of the chapter is an examination of how history and identity can affect a country's decision to go to war as well as examining the concept of distinct strategic cultures. Chapter 4 primarily focuses on international law and morality. The chapter reviews various attempts to regulate or outlaw war through international law, such as the League of Nations or the United Nations framework. Next the chapter engages with the idea of morality in an international relations context. While this book primarily adopts a realist framework, which is often characterized as amoral, the chapter argues that in an international system governed by imperfect information reputation is an important consideration, offering a space for morality in realist discourse. Chapter 4 concludes with a short examination of the role of nuclear weapons and whether war is even possible in such a context.

Part II focuses on various theoretical arguments loosely tied together by the idea that war is obsolete. This is divided into three chapters. Chapter 5 looks at theories that argue that war is fundamentally self-defeating. The chapter engages with two main theoretical schools of thought. First, it engages with defensive structural realism through Kenneth Waltz's *Theory of International Politics*. The key thesis of the chapter that defensive balance of power politics, as proposed by Waltz, has been interpreted as meaning that war is largely irrational since an expansionist policy would lead to the creation of a balancing coalition. Ultimately, it is argued that a policy of aggression would lead to defeat. It is important to note that this is not necessarily what Waltz has originally argued, rather than how the argument is often taken to its seemingly logical conclusion, especially in student discussions. The confusion often begins with the term "defensive", interpreted as a signifier for a theory contrary to the ideas of expansion. Chapter 5 first seeks to review the theoretical foundations of realism through Hans Morgenthau and then to review the key tenets of Waltz's defensive realism. This is necessary as the difficult language and detailed structural models of the original *Theory of International Politics* has proven to be a significant barrier to entry to students and an effective deterrent to the use of theory, favouring John Mearsheimer's offensive structural realism not necessarily for its theoretical correctness but easier readability. The chapter then demonstrates through examples selected primarily from the Asian theatre some points of criticism concerning the idea that defensive structural realism would consider all wars self-defeating and thus irrational. The second part of the chapter offers a similar treatment to the economic interdependence school of thought, using Norman Angell's *The Great Illusion* as its launching point. The chapter seeks to demonstrate that while economic interdependence can be an important deterrent to conflict, it is subject to power relations and is not an absolute prohibition on war.

Chapter 6 looks at theories that seek to offer alternate foreign policy options. It is important to note that the authors of these theories specifically state that they are not substitutes for military power. However, just as in the case of Waltz's defensive realism, these theories have often been interpreted as such, especially by new initiates to International Relations scholarship. The chapter reviews geo-economic statecraft through Blackwil and Harris's *War by Other Means* and soft power and influence through Nye's *Smart Power*. The arguments for both theories are similar: They offer compelling arguments for why military power is not an optimal tool under for certain scenarios, but they do not articulate a stance which would render war obsolete. War is not a universal tool and should never be considered as a first response. It is a niche tool to resolve problems that have proven to be resistant to other resolution methods. If anything, hard power and the capacity for war complement the tools discussed in this chapter to offer states foreign policy flexibility and the potential for synergy between different tools.

Chapter 7 takes arguments concerning the obsolescence of war a step further. The theories discussed in the previous chapters have focused on why it would be irrational to resort to war. The theories in Chapter 7 focus on the idea that the nature of international relations can be fundamentally different from a realist competitive model. The argument here is not so much about states considering but then rationally deciding against war, but rather war never even occurring to states. Mueller's *Retreat from Doomsday* argues that, reflecting on the horrors of war in the 20th century, developed states have come to the conclusion that war is unacceptable. In this argument, no developed state would consider going to war with each other or view war as a legitimate tool of dispute resolution. This book argues that a period of peace should not be confused with the obsolescence of war. Pragmatically speaking, rational calculations might have put a pause to the use of war, at least in the developed world, but that is not a rejection of war as a concept and with the alteration of strategic calculations the potential is still present to go to war. Wendt's constructivist interpretation similarly offers the possibility that through the evolution of identities a non-competitive international order could be created. However, this remains a theoretical possibility with limited pragmatic utility. After thousands of years of bloodshed, and in light of the competitive identities it has entrenched, even Wendt argues that it might be too late to hope for such an evolution. One may consider the example of China and Japan, where competitive identities are not only entrenched but are fundamental to both national identity and regime survival. At the end of the day it can make effective domestic politics to have enemies.

After establishing in Part II that the reports of war's obsolescence are greatly exaggerated, Part III moves into exploring the reasons as to why states would consider going to war. Once again, this part is divided into three chapters, organizing the wide range of reasons into three major themes: Systemic factors (Chapter 8), national characteristics (Chapter 9), and other practical considerations (Chapter 10). A war can hardly be traced back to a singular origin or reason. In most instances the decision to go to war is based on factors from all of these categories. For example, practical considerations such as the escalation of territorial disputes can explain the specific outbreak of hostilities, lighting the fuse, but they are dependent on larger structural and national factors to create an environment where there is a powder keg in the first place. Chapter 8 focuses on considerations that relate to the nature of the international system. The role balance of power considerations and security seeking behaviour plays in creating the conditions for war are explored through Mearsheimer's *Tragedy of Great Power Politics* and offensive structural realism. The chapter also examines the role of globalization and global rule setting as a setting for war by critically analysing Barnett's "core-gap" thesis that views globalization as an instrument of US power. Chapter 9 examines a set of three characteristics.

The geo-politics section looks at how the physical geography of Earth and deviations between this physical map and the political map that determines borders can be a motivating factor to go to war. The culture section uses Huntington's *Clash of Civilization* as a jumping point, although rather than examining civilizational conflicts, it looks at how culture and identity can be a significant component amongst the reasons for a war to occur. Finally, the chapter looks at domestic political organization, examining the claims of democratic peace theory in order to analyse whether domestic political organization matters to conflict initiation and whether that is a relevant consideration in an Asian context. Chapter 10 looks at issues that can specifically lead to the outbreak of a war. The chapter begins by analysing the escalation of territorial disputes as an ignition point to war, primarily focusing on Fravel's theory of perceived shifts in bargaining power. The second part of the chapter is dedicated to examining the idea of a diversionary war, that is, triggering an armed confrontation for domestic gains. The main purpose of Part III is to present a diverse picture on the reasons that could lead to a war to move away from a reductionist interpretation that it is simply the value of the territory or security seeking on its own that is responsible for the occurrence of an armed confrontation.

Part IV is dedicated to policy formulation. The previous two parts establish that war remains a tool within the state's foreign policy arsenal and examine the conditions that might lead to the occurrence of a war. However, the state still needs to make a conscious decision to "pull the trigger" on going to war. Part IV examines the decision-making process that could lead to such an outcome. This part adopts a rational choice model as its core, which is useful for both normative/anticipatory and explanatory research. The four chapters in this part take the readers through the different phases of the rational choice process. Chapter 11 focuses on the idea of rationality and the concept of bounded rationality as a foundation to utilizing this model. It also describes in detail how a rational choice model is constructed. Chapter 12 focuses on examining the feasibility of war. For a rational decision to go to war to be made, one first needs to establish that it is possible to accomplish the objective through going to war. At first look this might appear simple but determining the feasibility of war comes down to more than an aggregate total of national military power. This chapter provides a critical framework on how to make such a determination. Assuming that war is deemed feasible, Chapter 13 analyses the costs, benefits, and risks of going to war, using a diverse range of historical examples to illustrate the point. The rational choice selection of an optimal course of action is done based on such a cost–benefit–risk ratio. Finally, Chapter 14 examines alternate foreign policy options (coercion, peaceful resolution options, and delaying) to develop their respective cost–benefit–risk ratio, which serves as the basis of comparison in determining the optimal course of action. The chapter concludes with the process of determining the optimal course of action and taking the model further into a multi-level game theory context.

Notes

1 Mueller, J. (1990) *Retreat from Doomsday: The Obsolescence of Major War*. Basic Books: New York.
2 Pinker, S. (2011) *The Better Angels of Our Nature: A History of Violence and Humanity*. Penguin: New York.
3 Braumoeller, B. F. (2019) *Only the Dead: The Persistence of War in the Modern Age*. Oxford University Press: Oxford.
4 Mueller, J. (1990) *Retreat from Doomsday: The Obsolescence of Major War*. Basic Books: New York.

Part I

The nature of war

While the next two parts engage with the theoretical frameworks of war, Part I focuses on providing a solid foundation to that analysis by examining war as a concept. Chapter 2 starts by describing war for the purposes of this book, building towards a definition that is relevant for international relations and in a decision-making context. Chapter 3 focuses on examining the ideational context of war, taking a look at both on how history and identity can affect a state's approach to war, as well as states possessing potentially distinct strategic cultures. Chapter 4 looks at the legal and moral contexts of war. The international society has long tried to regulate war and this chapter will review these efforts from a historical perspective. The chapter also takes a critical look at morality in the context of war, arguing that even in a realist interpretation there is space for moral behaviour, even if from a point of self-interest. The chapter concludes with a brief examination of nuclear weapons and how they affect war in the contemporary international system.

The main point of Part I is to examine war in a number of different lights as it is tempting to view war from a single, reductionist perspective. That perspective is often that of a tragedy or moral evil. However, to critically examine the position of war in international politics, one needs to understand war as the complex activity it is. It is a tool of statecraft that is grounded in a number of contexts that affect how it is employed by various countries. This is especially so if the arguments of this book are used in an anticipatory fashion. Then it is crucial to understand how these different contexts translate across nations to maintain the validity of the analysis and avoid the trap of imposing one's perspective on distinct actors. Thus, while this part provides a general definition for war, it is also important to recognize that culture, history, and identity all affect the use of war as a foreign policy tool by nations. Nations such as China not only possess a distinct strategic culture that informs their decision-making and tactical choices during the course of a war, but also possess unique identities that inform their interests, their political decision-making, and their fears and willingness to go to war. This chapter explores all of these factors in order to create a strong foundation for critical examinations in the chapters above and for the creation of a rational decision-making model in Part IV.

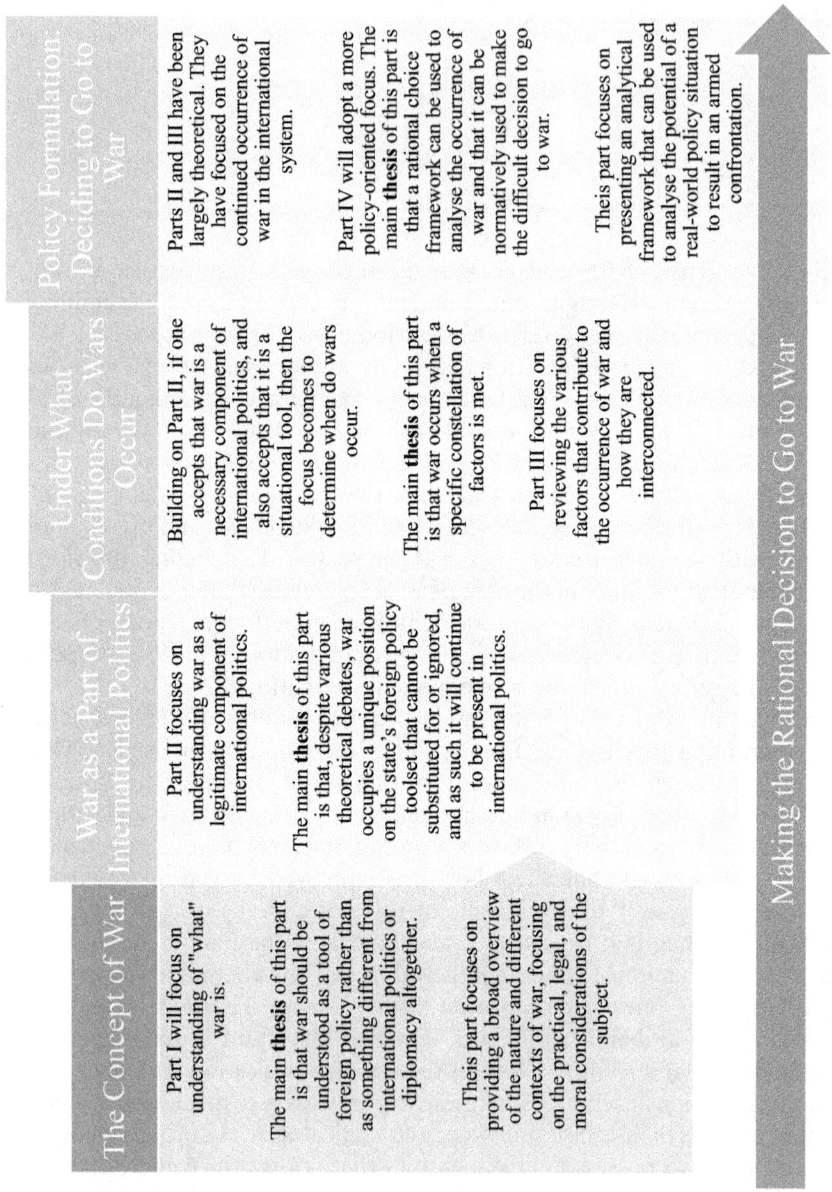

Figure PI.1 Part I's place in the overall arch

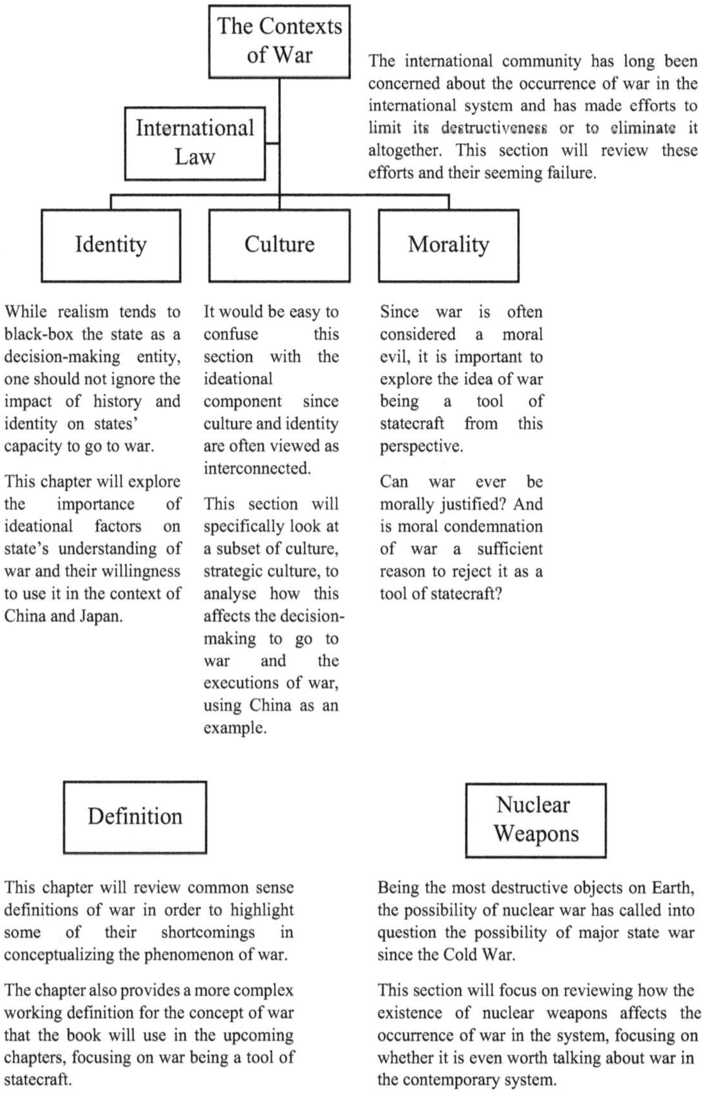

Figure PI.2 Structure of Part I

Table PI.1 Key works for Part I

	Author	*Work*	*Significance*
Chapter 2	Colin S. Gray	Fighting Talk: Forty Maxims on War, Peace and Strategy	*Fighting Talk* is a straightforward and succinct analysis on the nature of war, envisioning it as a tool of statecraft. It examines various levels of military power and how they play a role in international politics, arguing for the importance of proper strategy in connecting political goals with tactical military action.
Chapter 6	Michael Walzer	Just and Unjust Wars: A Moral Argument with Historical Illustrations	The book provides a critical and sometimes challenging overview of the morality of war. Is it acceptable to shoot sleeping soldiers? Why is war viewed as such a moral evil? Who is responsible for the crime of war? Do we have a responsibility to protect enemy combatants? These are all difficult questions the book seeks to tackle. The book forms the foundation for this book's interpretation of the morality of war, adopting Walzer's framework of *jus ad bellum* and *jus in bello* considerations.
Chapter 7	Thomas C. Schelling	Arms and Influence	Key thinker on the ideas of coercion and deterrence, describing it through the concept of diplomacy of violence where destruction serves as a bargaining chip. *Arms and Influence* is important to appreciate the multiple modalities in which military power can be used to achieve political objectives.

2 The definition of war

"What is war?" is one of those questions that seemingly has a self-evident answer. On the surface it seems to require no deep thinking or critical engagement to answer it. But as this section shows, in an International Relations context the answer is less straightforward than one might expect. "What is war?" is a fundamental definitional question for this book as this will underpin the discussion in the following parts. This chapter looks at various approaches to attempt to define the various conceptual meanings of war and coalesce them into the operational definition this book uses for the concept of war.

Basic definition

If one were to ask a person on the street at random to define "war", one would observe two common themes in their responses. On the one hand, descriptions would focus on describing war as a series of battles fought between soldiers and weaponry. If one would want to visualize this perspective, one can recall news footage of the Iraq War or one of the myriads of documentaries made on World War II. They tend to focus on soldiers and armaments engaged in combat to overcome the enemy in key battles. On the other hand, another set of descriptions would highlight the tragedy of war. War is commonly accepted as being hellish: A domain of violence, destruction, and death. Again, if one would want to visualize this approach, one could draw on the Vietnam War era picture of "napalm girl", aerial photographs of the bombing of Dresden, or images of the "highway of death" during the 1991 Gulf War. While both of these approaches offer indeed vivid depictions of war, this book would make the argument that they fundamentally fail to capture the functional nature of war, especially from a perspective relevant to international relations. These descriptions capture a specific component of war. But this focus on a specific part also tends to obfuscate the larger picture of war.

One can begin revealing the true nature of war by critically engaging with the first set of descriptors. Whether one focuses on battles, soldiers, and/or weaponry, one is capturing the *method* of war. Soldiers, fighter jets, destroyers, or submarines are weapons of war. They are not war in themselves. Similarly, battles are an integral part of war. These weapons of war are employed on the battlefield in the context of battles. It is often tempting to envision war as a series of these battles. However, again, this would be an inaccurate description. Fighting battles is meaningfully different from waging war.[1] Winning battles is often a method of war (although in some specific examples even that might not be required). But being good at fighting battles is insufficient to be successful at war. Gray points to the example of Germany. The Germans have been arguably very good at fighting battles. However, despite their battlefield performance and often advanced weaponry, Germany has suffered defeat in not one, but two world wars.[2] Based on this example alone, it is becoming obvious that such a description fails to capture something fundamental about war. Understanding how a war is fought (in the form of battles fought by soldiers and weaponry) is an important piece of the puzzle, but the whole picture will need additional pieces.

Looking at the tragedy of war provides an equally flawed description of war, once again capturing a piece of the puzzle only. Death and destruction are a *consequence* of war. But in most instances, they are not the reason for a war occurring, nor is it theoretically impossible to fight a war with little to no death and/or destruction. Walzer recalls the example of old African tribal conflicts, where, due to the customs of war, death was relatively infrequent because of the possibility of withdrawal from the battlefield.[3] Similarly, modern surgical strike capabilities can eliminate much of the destruction of war. Moving into the cyber realm, potentially one could paralyze an opponent, and thus theoretically win a conflict, without firing a single shot. Albeit in a different context, Sun Tzu argued that such bloodless conflicts, where the opponent loses before a single battle takes place due to it being manoeuvred into a position where resistance is impossible or irrational, are the true measure of skill for a general.[4] Thus, one can see that, just as in the case of the method of war, the consequence of war is an insufficient descriptor. Once again, one captures a piece of the puzzle. Combining it with the existing pieces can improve the quality of the picture, but still fails to capture its fundamental essence.

Both of these descriptions are of limited use because they fail to capture the *purpose* of war. Taking a quick glance at the question, the answer seems self-evident. The purpose of war is to defeat the opponent. This presents an elegant solution as it combines the method of war with the consequence of war, as defined before. Satisfied with the answer one is confident to sit back and relax. Unfortunately, based on such thinking, as a policy maker, one is likely to have dispatched their own armed forces on a military quagmire that will likely rival the United States' wars in Iraq and Afghanistan. Just as the previous considerations, "defeating the enemy" as a description of war fails to capture its fundamental nature. Equating war with the defeat of an opponent is an extremely common mistake, usually with grave strategic consequences. It suffices to say that there is a lot more to war than defeating an enemy through a series of battles with the use of soldiers and weaponry, suffering the tragedy of death and destruction in the process.

For the purposes of this book, one can start with Gray's definition of war and explore its different components in order to build a more complete picture on the nature of war. Gray defines war as organized violence for political purposes.[5] The violence part of this definition is the simplest to understand. It partially connects up with the focus on destruction discussed above in the context of popular descriptions of war. It is captured by both the battle-focused and destruction-focused descriptions. War requires the application of force to overcome the resistance of an opponent. Military power can accomplish this in multiple ways (which will be discussed later on), but whether it is applied directly or coercively, violence nevertheless remains a characteristic of it, incidental or otherwise.

The organized component of the definition focuses on the idea that this violence is undertaken with a purpose. War is not coincidental violence. It does not occur randomly. The state has to make a conscious decision to engage in war and then organize a military force to carry out this violence. Troops are regimented to a certain degree and carry out the will of the state. Even if one thinks of medieval peasant armies, there is an organization and chain of command to them. And they fight for a defined purpose. In the absence of organization, one would not talk about war. Largescale, non-organized violence could be visualized as rioting or ransacking a captured city. These examples would not qualify as war for multiple reasons. But in this instance, they would not qualify as it is not an organized form of violence. There is no clear structure to the forces and there is no translation of objectives into force through a chain of command. Large-scale violence occurs as an aggregate, but it is usually driven by individual or small-group actions, without an overarching guiding intelligence. Similarly, accidental engagements, such as cross-border fire, would not be considered examples of war, as they once again lack an overall purposeful organization to them.

The most important part of this definition is that war is fought for a political purpose. This is fundamental to understanding the true nature of war. One of the main shortcomings of the attempts discussed before has been their failure to capture this *purpose* of war, by focusing on its methods and consequences. War is only considered to be war if it is fought for the accomplishment of a political goal. That political goal can vary. It can be the capture of territory. It can be resistance against an invasion. It can be the unconditional surrender of an opponent. It can be the capture of sacred religious sites as seen in the case of the crusades. It can be to force a policy change. Whatever it is, for war to be war, it needs to have a political purpose. In the absence of such political purpose, one is not talking about war. It is possible that one is looking at criminal violence. One can refer back to the examples of rioting and ransacking cities above. A ransacking of a city can occur during a war. And if it is done as part of a coercive strategy as defined by Schelling (discussed later on), then it can be even considered to be part of "war" in a general sense, although this book will seek to separate direct use of force and coercion, considering the latter as distinct from "war" and as its own category of political violence.

Overall, by definition (for the purposes of this book), war is purposeful and organized violence. Before exploring the nature of war in detail, it would be useful to add one more component to the definition: It is interstate violence. When discussing war in this book, we will look at interstate conflicts, meaning conflicts fought between states. This limits the scope of conflicts that fit under the analytical model. It is useful to set up this distinction between interstate conflicts (conflicts between two states) and other conflicts because the theoretical foundations examined later, such as realism, often presuppose states to be the primary unit of analysis. Other forms of organized political violence can be grouped under non-state violence. This category would contain such instances as terrorism (political violence committed by non-state actors), civil wars (organized political violence within a state by different factions), state-sponsored or state terrorism (acts of political violence carried out by the state or related non-state actors at home or abroad), and genocide or ethnic cleansing (political violence often carried out by state actors against their own population). These are all important and troubling categories of violence. However, they are meaningfully distinct from the type of violence discussed in this book. This separation will be especially important when discussing war in the context of the international system.

The seven contexts of war

The section above provides a basic definition for war as organized violence committed for a political purpose. But Gray proposes that to truly understand war, one needs to understand it from the perspective of seven different contexts.[6] The political context of war has been discussed above in some detail already. As stated, for war to be war, it is always fought with a political objective in mind. This is going to be central to discussing war in the context of policy choices in Part IV. It is also integral to the basic definition of war proposed above.

The **socio-cultural** context is important because war is fought by societies with their own distinct cultures and values. One of the central themes of the rational choice analysis contained in this book is that identity is a key modifier affecting all levels of analysis as one has to examine decisions from the perspective of bounded rationality. Different societies view war differently. This has a few key implications for the analysis in this book. The most important ones are that (a) different societies afford different roles to war in their policy formulation process and (b) different societies develop distinct

Table 2.1 Seven contexts of war

Political	Military-Strategic	Historical	Technological
Socio-Cultural	Geopolitical & Geo-Strategic	Economic	

strategic cultures affecting both when force is acceptable and how force is utilized. This might not seem important from a normative perspective, but if one wishes to engage in a multilevel analysis that involves attempting to predict the moves of opposing states, then understanding the sociocultural contexts of war, especially how they affect state identities, will be of great importance. Potential differences in identities and strategic culture will be discussed in Chapter 3.

Wars always have "economic" and "technological" dimensions. Wars are economically taxing and often require a strong industrial and commercial base to pursue. Fielding a strong military is a monumental undertaking, usually reserved for major powers. This is especially so as a standing army is usually not a productive component of society. At the same time, war also results in economic devastation. This has to be considered if one wants to pursue war successfully. This was highlighted by Sun Tzu when arguing that a protracted war should be avoided as an economically exhausted state, even if victorious, is easy prey to others.[7] But beyond the economic costs of war and war preparations, there is a dimension in which war and economics intertwine. This was recognized by both the American naval theorist Alfred Thayer Mahan[8] and the Soviet admiral Sergey Gorshkov.[9] They both highlighted that naval military power exists to support one's own economy on land and to attack the opponent's economy during the course of a war. The use of military power was contextualized in its ability to bring a change to the economic balance of power. Industry and commerce underpin the ability of the state to fight. Disrupting it is a reliable way to diminish the opponent's capacity to resist – hence, the effectiveness of blockades, commerce interdiction, and scorched-earth tactics. The Soviet Union's ability to relocate industry out of harm's way during the German invasion was just as important as Russia's willingness to engage in scorched earth tactics to starve Napoleon's troops in preventing a successful invasion of the country. Similarly, to the economic context, and often interconnected with it, is the technological context of war. The connection is not hard to understand: The economically stronger a state, the better weaponry it can afford. But technology underpins how one fights wars and thus whether one can successfully pursue war as a policy course. The introduction of air power, coal-powered warships, breech-loading rifles, or gunpowder itself has fundamentally affected the battlefield and war in general. And when one's method of warfighting has not caught up to technological advances, one can witness some of the bloodiest conflicts in history. Relying on human wave tactics in the age of machine guns and artillery led to mass casualties in World War I and preceding conflicts such as the Russo-Japanese War of 1915. Similarly, outdated tactics in the face of improving rifle and artillery technologies contributed to the death toll of the American Civil War. When studying war, one needs to critically examine the particular technological context in which a war was conducted as this can contribute greatly to understand the outcome of the conflict. While there is a risk to overestimating the importance of technology over the human element, adopting a purely technocratic approach to war, superiorly equipped forces tend to have an advantage on the battlefield.

The "military-strategic" context refers to the actual availability and use of military power. Balance (or imbalance) of power is a central component to realist analysis of International Relations. How much military power one possesses relative to the opponent is a fundamental consideration for the study of war. This book will depart from the traditional focus on total military power observed in a large part of International Relations scholarship (due to its simplicity) to favour a focus on a contextual approach to balance of power. To understand one's potential to successfully pursue a military conflict, one needs to be clear on the military power needed to do so in a specific scenario and how to best employ it in a strategic sense on the battlefield to accomplish the overarching political objective. The balance of power will be discussed in detail in Part IV, but it will be periodically brought up when examining different theories relating to the conduct of war in the international system in Parts II and III. It is important to note that this is not a book of strategy or tactics. It does not seek to examine optimal battlefield strategy to accomplish objectives, instead favours a high-level view from an International Relations perspective. But basic understanding of the nature of the conflict one needs to fight in order to accomplish the desired political objective is important to determine the feasibility, costs, and risks of the enterprise.

The "geopolitical and geo-economic" context refers to the importance of physical geography in the conduct of war. This will be examined in detail in Chapter 9 when engaging with the work of Robert Kaplan. At this point it suffices to say that as long as one is constrained to a terrestrial environment, the physical formation of the planet will affect the conduct and objectives of war. Geographical advantages are going to be relevant both to the conduct of battles and as desirable political objectives for a conflict. While technological advancements offer the illusion of geographical nullification, one would be ill-advised to ignore geography as a factor. Despite air power, satellites and long-range missiles, rivers, mountains, and deserts continue to fundamentally affect our strategic thinking. For example, China's strategic behaviour cannot be understood without understanding the geographical vulnerability of the economically important coastal regions. Why China would desire to control small reefs and islets, even at the risk of an armed confrontation, can only be properly analysed in the context of the geography of the South and East China Seas, where these groups of small land formations create critical chokepoints and valuable bastions.

Finally, "history" offers an important component similar to the sociocultural context. Wendt argues that state identities are strongly influenced by the history of shared reciprocal interactions.[10] State identity will be an important consideration in understanding interstate behaviour in a bounded rationality context. For example, historical animosity and rivalry can be an important motivating factor for competition. Similarly, a history of invasion and military aggression can amplify threat perceptions and a sense of insecurity, leading to what Mearsheimer describes as "fear" between states that underpins the severity of competition in the system.[11] Ultimately, Sino-Japanese competition makes no sense without understanding their shared history and how it affects their current strategic behaviour. History is also important in a normative context. Past actions are an important, if imperfect, indicator of future behaviour. When engaging in anticipatory or predictive analysis, historical precedents and patterns of behaviour provide much needed guidance in an otherwise impenetrable fog of uncertainty.

Overall, one can observe that war is a complex subject. While war has captured the popular imagination, the nuance of the enterprise is often lost in the focus on its most visible attributes. To properly understand war, one has to be willing to bring in a certain level of interdisciplinarity, especially if looking at war from the perspective of the international system. There is much more to war than warfare. Often the most interesting conundrums of war lie beyond the battlefield, including understanding how a state can come to the rational decision to unleash the horrors of war, which is the primary subject of this book.

War as an instrument of statecraft

Looking beyond the definitions, this book proposes to examine war in the context of foreign policy, as this would be most relevant to International Relations scholarship. Definitions above have emphasized the political nature of war as a key defining feature of it. But it is common to view war as nevertheless distinct from politics. Von Moltke, the Prussian general, is quoted to have said that war begins where politics ends.[12] This is often quoted in support of the common argument that war in international relations is a phase distinct from politics. If anything, war is often viewed as the result of the breakdown of the political process, going to war being viewed as a failure of diplomacy. One could visualize this as the passing of leadership from politicians and diplomats to generals. This book would disagree with such an interpretation. One of the key theses of this book is that war is politics. It is neither distinct nor separated from it. And thus, war should be considered in the context of foreign policy, as a tool available to the state to use along with other instruments of statecraft.

Instead of von Moltke, one can look at the similarly popular quote from Clausewitz: "[W]ar is the continuation of politics by other means".[13] Clausewitz has been similarly interpreted to mean that politics and war are separate domains. However, a closer inspection of the quote, especially in the context of Clausewitz's own theories, reveals the opposite intended meaning. To put it into its proper context, one can take a different quote from *On War*: "[W]ar is not merely a political act, but also a real political instrument, a continuation of political commerce, a carrying out of the same by other means".[14] Clausewitz regards war as a political instrument. Revisiting the original quote one can discover the meaning of "continuation of politics by other means". Clausewitz does not regard war to be separate from politics. On the contrary, *On War* views war as part of the political process. The focus on *other means* simply translates into war being a tool in the state's larger arsenal.

If we carry on Clausewitz's thinking, then for the purposes of this book one can consider war to be an instrument of statecraft, not inherently distinct from other avenues of foreign policy. There is an immediate benefit to this: It resolves a logical inconsistency other interpretations suffer from. As discussed above, war is considered to be inherently political in nature. If war has a political purpose but it is considered to be distinct from politics, then one has to deal with a logical gap. If one goes with the interpretation of von Moltke that war begins where politics ends to mean a distinct separation between politics and war, then the insistence on the political nature of war makes little sense. Military necessity supersedes political considerations once the first shot has been fired. In contrast if one accepts war as part of the course of political behaviour in the international system, then no such inconsistency emerges and the insistence on the political nature of war makes perfect sense. War is merely a method to achieve political objectives the state has failed to accomplish through other avenues of diplomacy.

Such an interpretation might appear to be morally objectionable. It would most certainly seem to go against existing prohibitions on war in contemporary international law. War is terrible. It is often considered distinct from politics to denote that it is abnormal or unacceptable conduct in the international system. However, while such motivations are understandable, they obfuscate the nature of war by divorcing it form its inherent contexts. Wars are indeed terrible. Yet they keep occurring in the international system. There is no escaping that. If one hopes to limit their occurrence through the purposeful crafting of policy in what one can call preventative diplomacy, then war needs to be understood as a phenomenon of international relations without the fog of moral objection. War is a tool of statecraft, not inherently different in its purpose to other elements of statecraft such as diplomacy, economic engagement or geo-economic statecraft. It only differs in its execution and particular characteristics, not in its fundamental nature.

Thus, for the purposes of this book, war will be discussed in the context of the state's larger foreign policy agenda. From an analytical perspective, war will be considered as a component of the state's foreign policy arsenal, competing with other tools to offer the optimal course of action to rational decision-makers. It will be analysed with regard to its ability to deliver political objectives and in the context of its relative attractiveness to other avenues of foreign policy. This does not equate to a moral acceptance of war. However, International Relations is a field where a decision-maker is often presented with only flawed or bad choices, entrusted with the burden to nevertheless chart the policy direction of their nation. A rejection of war as a tool of foreign policy is understandable but the simple recognition that it is part of the same considerations does not equate to an unconditional support for states to go to war. However, an improper understanding based on moral preconceptions all but guarantees a failure to properly manage the occurrence of wars in the international system.

Different forms of violence in the international system: Use of force versus coercion

As a final consideration to the definition of war, it is important to note a definition specific to this book. For the purposes of this book, and the policy analysis model contained within it, war and military coercion will be considered to be separate avenues, despite the fact that they are often intertwined and sometimes indistinguishable from each other. This separation is made based on the model of Schelling, underpinned by the distinction in the methods of how military power is applied in a certain situation.

Schelling argues that military power can be applied in two major ways. The traditional approach can be described as use of force (or direct use of force). In this instance, force is used to seize and hold an objective directly, bypassing the enemy's will as a target, focusing simply on the military might of the opponent. Destruction and carnage are incidental to the undertaking. Schelling argues that this is distinct from coercion or "diplomacy of violence" (indirect use of force). In coercion, destruction is used to overcome the opponent's will to resist. Death and destruction in this scenario are purposeful. It is used as a bargaining chip. It is the opponent's fear of further damage that persuades them to concede. As such, coercion is inherently psychological in nature.[15]

The separation between the two can be difficult. Clausewitz himself seems to conflate the two. *On War* highlights that the target of military operations is the enemy troops to diminish the opponent's will to resist through strategic destruction. While the objectives of war in *On War* seem to closer correlate with what Schelling would describe as use of force, the method seems to offer a psychological component more akin to diplomacy of violence. This conflict can be resolved considering Clausewitz deeper. In Schelling's interpretation, in a use of force scenario the main obstacle to one's success is the opponent's military might. This has to be overcome to seize and then hold on to the desired objective. The destruction described by Clausewitz serves this point. The violence is not directed at the enemy psychologically to undermine their will to resist through the horror of war, which would be the domain of diplomacies of violence. This is partially supported by Clausewitz's focus on proportionality and balancing the destruction of the opponent to preserving one's own strength. Clausewitz is still describing a direct contest of military might, not a psychological manipulation of decision-makers through the threat of violence.

To illustrate this with an example, one can consider the cases of the 1974 Battle of Paracel Islands and the 1995 Mischief Reef incident. In both instances China has acquired contested territory through military might. Yet in each instance military might was applied in distinct ways. In the Battle of

Paracel Islands, there was a direct military confrontation between Chinese and South Vietnamese forces. China has accomplished a military victory by dispatching local enemy forces and held on to the islands through military superiority. Victory was achieved as South Vietnam calculated the costs of trying to recapture the islands and escalating tensions in the process to be higher than what the islands were worth. In contrast, in 1995 China has captured the Mischief Reef without firing a single shot. The Philippines has contested China's control over the islands, but the presence of military assets in the region has convinced Manila to back down. It was the threat of violence, the destruction the superior People's Liberation Army could inflict on Filipino forces, and the overall economic and other damages the Philippines would suffer in escalating the dispute against a more powerful opponent that persuaded the Philippines not to counter. Its forces were not defeated and China has only dispatched a nominal military force into the region. While South Vietnam has suffered a military defeat, the Philippines suffered a psychological defeat. While in execution the two conflicts might have appeared similar (students on examination often confuse use of force and coercion), the nature of the application of military force was distinct. This distinction is going to be especially important as coercion morphs into the "grey-zone challenges" one increasingly observes in territorial disputes in the Asia-Pacific region.

For the purposes of this book the separation will be made in order to focus on direct contests of might. When referring to war in the following chapters, it will be in reference to conflicts that would fall under such contests, and examples of diplomacy of violence will be denoted separately under the terms of "coercion". While this distinction might seem arbitrary, it serves an analytical purpose. The demands and execution of direct use of force and coercion are meaningfully different. Direct use of force focuses on the deployment of military power to seize objectives. Coercion, according to Schelling, is most effective if the violence is held in reserve. It is the threat of violence that should coerce the opponent to concede – or the threat of further violence in many instances, as credible threats often rely on demonstrative actions. In some instances, coercion might even require a full-scale traditional war to precede it to remove the opponent's ability to resist the damage one seeks to use as a threat.[16] But again, as they apply military power for different purposes, this book will consider them as separate policy avenues. This is especially so, as in a rational choice context, having the ability to clearly define the costs and benefits of competing actions is of the outmost importance. Combining coercion with use of force would muddy the water when it comes to defining key considerations for a rational policy decision.

To summarize a formal definition of war for the purposes of this book, war is considered to be interstate organized violence for political purposes, where force is used to seize and hold objectives, instead of bargaining through destruction and threats. War targets the opponent's military might rather than primarily seeking to psychologically undermine the opponent. One can see that this definition has evolved considerably over this chapter compared to its earliest form that envisioned war merely as a series of battles.

Levels of analysis: Policy, strategy, and tactics

When discussing war, it is important to distinguish between different levels of analysis. Confusing these levels is rather common and, as Gray points out, often leads to flawed execution.[17] In discussing war, one has to distinguish three key levels of analysis: Policy, strategy, and tactics. Each refers to a specific aspect of war. This book will offer specific definitions for each, although they may be used differently in the overall literature.

Tactics refer to the operational level consideration of a war. Tactics is a domain of warfighting and military personnel. Analysis on the tactical level focuses on the execution of the war itself and planning and fighting battles.

Strategy is an intermediate level between tactics and policy. While strategy focuses on the execution of the war, the main focus of strategy is to translate political goals into objectives that can be achieved on the battlefield. Strategy is not primarily concerned with military minutiae, but rather focuses on whether political change can be accomplished through military action.

Policy is the highest level of analysis. Policy focuses on the overall political objectives of the state. Policy is primarily of the domain of politicians and political scientists. Policy analysis mainly focuses on identifying the key interests of the state and determining the best tool to pursue them.

Gray argues that for the effective execution of a war, these three levels of analysis have to be in balance and need to cooperate effectively.[18] Breakdown on any level, or in translation from level to level, often results in a military disaster. Improper tactics could lead to battlefield defeat that makes winning a war improbable. Pursuing improper political objectives is similarly disastrous for the state. But equally important is translation. One can have appropriate political, strategic, and/or tactical objectives. But if there is no proper translation to all other levels, then the war will not be successful. A frequently quoted example is the United States' war in Iraq. While the Bush Administration had both clear political and military objectives, there was no direct connection between the two levels or any idea of how the accomplishment of military objectives would translate into political change.

This book will primarily focus on the policy level of analysis. The goal is to devise a decision-making framework that can explain the continued occurrence of war in the international system and offer a normative framework to reach the rational decision to go to war. To do so, one also has to gain some familiarity with the strategic level of analysis. Understanding a basic level of strategy is necessary to determine whether a policy could be pursued through the means of military power. Thus, such understanding is needed even if one focuses on the policy level. Traditional wisdom suggests that war is a domain of generals. However, policy makers cannot afford to be ignorant about the affairs of war if they seek to effectively wield it as an instrument of statecraft. This is especially so as military personnel, by function, tend to be preoccupied with the operational demands of war.

Overall, it is important to recognize how all three levels need to cooperate during a war. The decision to go to war is primarily a policy-level question. But without understanding the operational demands of war, one would be hard-pressed to make an informed decision. Part IV will explore this matter further.

Notes

1 Gray, C. S. (2009) *Fighting Talk: Forty Maxims on War, Peace and Strategy*. Potomac Books: Westport.
2 Ibid.
3 Walzer, M. (1977) *Just and Unjust Wars*. Basic Books: New York.
4 Sawyer, R. D. (2007) *The Seven Military Classics of China*. Basic Books: New York.
5 Gray, C. S. (2009) *Fighting Talk: Forty Maxims on War, Peace and Strategy*. Potomac Books: Westport.
6 Ibid.
7 Sawyer, R. D. (2007) *The Seven Military Classics of China*. Basic Books: New York.
8 Holmes, J. R., and Toshi Y. (2008) *Chinese Naval Strategy in the 21st Century: A Turn to Mahan*. Routledge: London, New York.
9 Groshkov, S. G. (1979) *The Sea Power of the State*. Pergamon Press: New York.
10 Wendt, A. (1992) Anarchy Is What States Make of It: The Social Construction of Power Politics. *International Organization* 46(2), pp. 391–325.
11 Mearsheimer, J. J. (2014) *Tragedy of Great Power Politics*. Norton: New York.
12 Gray, C. S. (2009) *Fighting Talk: Forty Maxims on War, Peace and Strategy*. Potomac Books: Westport.
13 Clausewitz, C. (1874) *On War (Translation by J.J. Graham)*. Guttenberg.
14 Ibid.
15 Schelling, T. (2008) *Arms and Influence*. Yale University Press: New Haven.
16 Ibid.
17 Gray, C. S. (2009) *Fighting Talk: Forty Maxims on War, Peace and Strategy*. Potomac Books: Westport.
18 Ibid.

3 War, identity, and culture

The question of whether states are differentiated by various characteristics is a fundamental conundrum of International Relations. Theories have long pondered whether states are functionally identical units in the international system or whether there are such things as "Asian way" or "European way" of politics. From a constructivist perspective, war plays a crucial role in the formation of national identities due to such conflicts' high-impact nature. Engaging in war can significantly alter the identity and behaviour of various states. Sticking with the examples of China and Japan, both of their identities have been significantly affected by their military conflicts. Furthermore, these conflicts continue to underpin their contemporary relationship. While realism provides an important foundation for International Relations, constructivism and its contributions on identity cannot be ignored. If one looks at the central argument of Wendt's constructivist interpretation of anarchy (discussed in Chapter 9), the main theme is that states interpret the world based on past experiences. A history of invasion and war is likely to make a country to interpret the world in competitive terms, especially if such experiences are reinforced through a political-nationalist narrative. Such a state would be expected to be highly concerned about territorial defence and would view the rest of the world with suspicion, more so than a country with no such history or a history of successful use of force. Similarly, legacies of being an aggressor in a conflict can create long-lasting effects on state identity and relationship towards military power.

China: Strong defence and historical victimhood

China's contemporary identity cannot be understood without examining its military history. Insecurity is a strong theme in contemporary Chinese political identity, which underpins China's assertive security-seeking behaviour in the international system. This is strongly rooted in China's military defeats in the late Qing period and the security dynamics of the Cold War. These influences underpin an identity that tends to interpret security in competitive terms and that is suspicious of the intentions of other states, conforming to a largely offensive structural realist model of interstate behaviour. However,

China's insecurity goes beyond that which would be justified by purely geopolitical considerations, hence highlighting the importance of identity.

Late Qing and Republican Period

The late Qing period has been characterized by the quasi-colonial subjugation of the Chinese Empire through a series of military defeats. The Chinese Empire has traditionally been a regional great power, but this has been undermined by Western imperial powers. The first of these military conflicts has been the 1839–1842 First Opium war. The war was the result of growing tensions between British economic interests, including the sale of opium of China to address a significant trade deficit with the self-reliant Middle Kingdom, and China's attempt to maintain sovereignty and internal control. The war ended with a Chinese military defeat, resulting in the signing of the Treaty of Nanking, the first of what China considers to be "unequal treaties". Unequal treaties are characterized by granting significant privileges to imperial powers. Most of the treaties were coerced as a conclusion of various military engagements.

This period of Chinese history is characterized by a string of military defeats, leading to the ever-increasing erosion of Chinese sovereignty. By the beginning of the 20th Century, China has been reduced to a colonial status, with various foreign powers enjoying significant concessions and having the ability to directly interfere with the running of the state. This was underpinned by the military weakness of China. The colonial subjugation was only possible because China lacked the ability to repeal foreign invaders. For example, China has traditionally focused on riverine defence, with little interest in the maritime realm. Consequently, China has been ill-equipped to deal with superior European naval power, which was leveraged into the dismantling of riverine defences, allowing European forces to penetrate the Chinese interior. While China attempted to reform its military forces in the Qing period, both army and navy, these proved to be too late to reverse the decline. This also created a vicious cycle: Military weakness led to more concessions and internal weakness, which in turn made military reform ever more difficult.[1]

The experiences of the late Qing period are replicated in modern Chinese patriotic discourse. These experiences support China's attitude that it has to survive in a hostile international system and that this necessitates strong territorial defence. Having been reduced from the Middle Kingdom, the centre of civilization, to the "sick man of Asia" has left deep marks on China's security perceptions. Events such as the burning of the Summer Palace are used as powerful symbols of the Chinese Communist Party's patriotic narrative that continued vigilance is necessary in a hostile world seeking to keep China down.[2] These experiences coalesce in what is called the "Century of Humiliation". It is a period of Chinese weakness, frequent warfare, and defeat.

These perceptions are exacerbated by the rise of Japan. Japan accomplished becoming the first, and only, Asian nation at the time to achieve parity with

Western powers. However, from a Chinese perspective, Japan, a former beneficiary of Chinese civilization, leveraged this into seeking similar colonial domination of China. By 1895, China not only did suffer defeat to more advanced Western imperial powers, but to an Asian power China traditionally viewed as a lesser. During the First Sino-Japanese War, Japan has defeated China militarily and acquired similar concession as Western imperial powers. During the 1900 Boxer Rebellion, Japan also participated as an equal in the Eight Nation Alliance against China. These further marked just how far China has fallen from its traditional position of regional leadership.

The Qing Dynasty came to an end in 1912, giving way to China's republican period. However, the Chinese Republic continued to suffer from weakness and foreign domination. The period is characterized by domestic division and the rule of warlords, as well as the early phase of the Chinese Civil War between nationalist Kuomintang and communist forces. Against this backdrop, the country faced the increasing assertiveness of the Japanese Empire. In 1915 Japan posed its Twenty-one Demands to China, requiring broad sweeping concessions, including appointments that would have virtually ended China's sovereignty. Following the 1931 Mukden Incident, Japan has created a puppet state in Manchuria, installing the last Qing emperor as its figurehead. Animosity between the two countries has escalated until Japan's full-scale invasion of 1937.[3] China's War of Resistance against Japanese Aggression has been characterized by unprecedented brutality and crimes against humanity. Just as the burning of the Summer Palace is a powerful symbol of the downfall of the Chinese Empire, the Nanking Massacre is used as a symbol of the dangers China has faced in the international system at the hands of foreign invaders.[4]

Ultimately, from an identity perspective, the experiences of the late Qing and republican periods have underpinned China's sense of insecurity and desire for strong territorial defence. This is disseminated and reinforced by China's post-1989 patriotic narrative. Using Wendt's terminology,[5] it is hardly surprising that China would view its relationship with Japan and Western powers as strongly competitive. Suffering and eventually overcoming the humiliation of foreign aggression is a central element of the Communist Party's mythology. War has played a central role in shaping China's contemporary identity.

Cold War period

While the Cold War period lacked large-scale military confrontations, the dynamics of the period nevertheless affected China's identity. After the conclusion of the Chinese Civil War, the communist having reunified the mainland sought to rebuild the country. They did so with Soviet assistance. This early period has been characterized by hostility between China and the West, especially as the United States backed the Chiang Kai-Shek's nationalist regime on Taiwan, including securing China's contested membership in international institutions such as the UN. Mao's revolutionary ideology focused on world revolution and the constant threat posed by the US-led capitalist block. The United States met China on the battlefield during the Korean War, when China intervened on behalf of North Korea alarmed by McArthur's advance towards the Yalu River. Even after the war, animosity remained constant, with the United States eventually seeking the use of nuclear weapons against China, deterred by the Soviet Union.

However, China witnessed some of the most dramatic turnarounds during the Cold War. By the 1970s, the relationship between China and the Soviet Union drastically soured. While the Soviet Union was essential in providing assistance to China in the 1950s, by the 1960s disagreements led to the Sino-Soviet split. The Soviet Union increasingly regarded the PRC as a threat, going as far as seeking US support for the use of nuclear weapons to halt China's indigenous nuclear programme. China is one of the few countries that was threatened by both the United States and the USSR with nuclear war. Sino-Soviet animosity led to the PRC switching sides following the meeting between Mao and Nixon in 1972.

The Cold War period continued China's sense of geopolitical insecurity. Ideological commitments or alliances were determined to be largely meaningless. The Cold War experience of China has been a confirmation of the interest-driven nature of the system, where there are no permanent friends or enemies, only permanent interests. It is hardly surprising that realism, including offensive structural realism, continues to have a strong influence on Chinese strategic thinking. China sees in its own historical experience the replication of the competitive tenets of the realist school of thought, including a constant pressure for competitive behaviour.

This period has also been characterized by offensive use of force. China has resorted to the use of force against UN forces in the Korean war (1950–1953). The early history of the PRC is characterized by an ongoing campaign against nationalist forces in littoral waters, where China sought to secure control over various islands and to counter nationalist attacks. In 1962 China engaged in a military confrontation with India over disputed territory in the Himalayan border region. In 1969 China and the Soviet Union engaged in a border skirmish over the sovereignty of islands in the Ussuri River. In 1974 China engaged in a naval skirmish with South Vietnam over the sovereignty of the Paracel islands. Similarly, in 1979 China went to war against Vietnam, motivated by a desire to counter Vietnam's own regional ambitions and the increasingly threatening Soviet presence in the region. The use of force remained an important component of the PRC's foreign policy toolbox.

Identity and China's contemporary security posture

The security posture of contemporary China is underpinned by a deep sense of insecurity. Demographic and economic changes in the post-Mao necessitated the re-evaluation of the PRC's strategic priorities. China's economic concentration in the coastal regions, combined with a weak presence in a saturated maritime environment, creates a tangible insecurity for China. However, China's sense of insecurity goes beyond these considerations. While the Chinese coast is vulnerable, an attack against China is not imminent. Regional rivals lack the amphibious and other capabilities to significantly threaten China. Japan certainly has shown no interest to project military power into mainland China. The United States has been identifying China as a near-tier competitor, awarding it a central focus in its defence planning. But, similarly, there is no indication that an armed confrontation with the United States is imminent or that China's nuclear deterrent would have lost its effectiveness.

Yet, China has been engaging in assertive security-seeking behaviour, even at the cost of its regional reputation. China has been reliant on coercion to pursue its interests over the East and South China Seas, confirming the very "China threat" theories previous Chinese administrations sought to alleviate. Much of the efforts of Hu Jintao to portray China as a "good neighbour" in the early 2010s[6] has been expended. Xi Jinping's China is viewed with considerable concern. This is most evident in the strategic posture of Japan. While the right-wing Liberal Democratic Party of Japan has always sought to alter Japan's security posture, China provided the necessary political capital to Shinzo Abe to meaningfully reorganize Japan's defence posture. While these reforms fall short of China's claims of remilitarization, they represent an increased focus on deterrence in Japan's security posture over its traditional emphasis on economic engagement.

China's security-seeking behaviour is imposing significant costs on its regional diplomacy. Deng Xiaoping cautioned against such behaviour when arguing that China should bide its time and keep a low profile in order to avoid triggering balancing action. The intensity of China's behaviour can be especially puzzling to explain when one considers that war is not imminent. If anything, China's own security-seeking behaviour is pushing the region closer to a military confrontation, realizing realist security dilemma dynamics. While the explanation for such a behaviour is multifaceted – there are significant domestic political considerations relating to the legitimacy of the Chinese Communist Party in the absence of democratic consensus – identity is an important component of it. Through the history shortly described above one can see that victimhood is a strong component of Chinese history. This motivates China to assess its security situation as more severe than it might realistically be, creating stronger incentives for security-seeking behaviour. War and humiliation being such central elements of Chinese political narratives affects Chinese identity in a way that predisposes the country to assess its relationships in competitive terms and to emphasize the need for strong national defence.

This is especially true in the context of Japan. China views Japan in strongly competitive terms based on their shared experience of war and the brutality of Japanese invasion. Japan has attempted to alter this relationship through engagement and economic assistance. But the positive experiences proved far less significant in shaping the relationship than the negative experiences of war. This highlights the importance of war in shaping the identities of nations relative to others. Again, this has a political dimension of securing the legitimacy of the Chinese Communist Party. But it is undeniable that their shared history of war plays a key role in the relationship until today, despite the fact that post-war Japan sought to provide substantial development assistance to China and that there is significant economic interconnectedness between the two countries.

Essentially, the Century of Humiliation period (1842–1949) and Cold War dynamics created a strongly insecure Chinese identity where only becoming a great power is viewed as a position of security. This creates a revisionist power that often pursues assertive policy stances or reacts strongly to otherwise mundane international incidents. Based on such an identity, no position short of regional dominance would be viewed as secure, creating a growingly tense relationship with other regional states.

Japan: From militarism to anti-militarism

Japan presents an interesting example for the interconnectedness of national identity and military power. The customs of the classical Japanese warrior class have intertwined greatly with its societal values, but at the same time modern Japanese society is often characterized by a strong sense of anti-militarism and opposition to military power. Examining the history of Japan, especially how society approached military power, can highlight how identities evolve in the international system.

The evolution of the samurai class

Despite their antagonistic relationship, Japanese and Chinese history offers several similarities. One of these similarities is that both countries have had important warring states periods. National unification under a single dynastic leadership has been an important element of both of their histories, whether one is talking about the Chinese warring states period (5th century BCE–2nd century BCE) and the Romance of the Three Kingdoms period, or Japan's own warring states period. The *sengoku jidai* period (1467–1600) preceded the unification of Japan under the Tokugawa Shogunate.

The rule of the Tokugawa Shogunate represents roughly 300 years of Japanese history that created the foundations of contemporary Japan as a state. As a primarily feudal system, the warrior class has played an important role in society and its own traditions have diffused into popular customs. Samurai values have been incorporated into the official culture of the Shogunate. The Tokugawa Shogunate emphasized social order and orthodoxy, creating a rigid framework for society. This period is well characterized by the fact that Tokugawa Ieyashu defined rude behaviour – for which a samurai could institute capital punishment – as acting in a manner other than expected.[7] The Tokugawa Shogunate has sustained its rule through controlling all aspects of Japanese life.

In this rigid system, the samurai played a key role. The samurai as a warrior class often traces its history back to the Mongol invasions of Japan in 1274 and 1281, where Japanese warriors have defeated the superior Mongol forces. In the feudal society of Japan, similarly to European states, the warrior class did not only serve in a military function but took on various administrative roles. The samurai were responsible for enforcing the social order day-to-day. However, unlike in Europe, the Tokugawa Shogunate pursued a policy of isolationism. During the *sakoku jidai*, in the absence of foreign threats and enjoying relative domestic stability, the samurai gradually transitioned during the Tokugawa Shogunate to become primarily an administrator class, with their importance as a military force diminishing.[8]

A significant source of diffusion of warrior customs was the creation of the *bushido* code, a set of social norms practised by the samurai class. Being so closely associated with Japanese society, one could be forgiven for not realising just how recent the term is. While the values of the samurai influenced Japanese society throughout the Edo period of Japan (1603–1868), bushido was largely codified by Inazo Nitobe's 1899 book, *Bushido: The Soul of Japan*. Many contemporary values, such as a focus on honour and social standing, have a strong connection to the bushido code. Just as some components of medieval chivalry have diffused from the knights (warrior class) to society in general, so did the customs of the samurai class.

It is interesting that the ideas of bushido became commonplace at a period when the samurai class has declined and eventually disappeared altogether. The intrusion of Western imperialism has critically weakened the Tokugawa Shogunate, leading to the dismantlement of traditional social order during the Meiji reform period. The samurai class weakened as the Meiji reform period pursued the Westernization of the country, including increased social mobility and the dismantlement of the feudal administrative system.[9] The samurai class met its end during the 1877 Seinan Rebellion where Japanese imperial forces have defeated the remnants of the samurai class.

Pre-war Japan

The pre-World War II history of Japan is a complicated period characterized by contradictory political forces. The Meiji reform period and the *Taisho* period brought the Westernization of Japan, including increased social mobility, the abolishment of the rigid class system, and democratization. But his period has also been characterized by the rise of nationalism and autocratic tendencies. Japan sought to emulate Western powers to avoid the quasi-colonial fate of China. However, imitating Western powers meant also internalizing competitive security behaviour and imperialism. During this period Japan has embarked on its own imperial project, including the annexation of Korea and

Taiwan, the 1st Sino-Japanese War of 1895, participating in the Eight Nation Alliance to put down the Boxer Rebellion and to secure concessions in China, and the Russo-Japanese War of 1905.

During the Meiji reform period Japan exchanged its traditional feudal army of samurais for a contemporary conscripted army modelled on Western powers. While Japan was modernizing on all fronts, military modernization specifically was a large portion of government expenditure. This was a general trend in the pre-war period, leading to arms control conferences, for example, on naval extension such as the 1922 Washington and 1930 London naval treaties. The burden of military expansion was compounded by the difficult economic situation of Japan after its export boom during World War I has dried up and the effects of the Great Depression were becoming more prominent. In Japan's contradictory political climate, the military sought to establish its own political voice and to secure its own political designs. In the pre-war period the army and navy gained increasing political importance, on occasion capturing the office of the prime minister among Japan's rapidly rotating prime ministers (between 1900 and 1940, Japan had 30 prime ministers).

During the 1931 Mukden Incident, the Kwantung Army acted without civilian authorization to invade Manchuria, under the pretence of responding to sabotage against the key South Manchurian Railway. This marked the point where the Japanese armed forced essentially broke free of civilian government oversight and begun acting on their own initiative.[10] Then Prime Minister Inukai Tsuyoshi resisted the army's efforts to incorporate Manchuria into the Japanese Empire through the puppet state of Manchukuo. He was assassinated by naval officers in 1932, marking an end to civilian control in Japan. The Japanese state, dissatisfied with its regional position, sought closer ties with other revisionist powers such as Germany and Italy, instead of its World War I ally Britain and other Allied powers. In 1937 Japan begun the invasion of China, marking the opening of World War II on the Pacific Front.[11]

Post-war Japan

World War II ended up as a traumatic experience for Japan. It lost the conflict leading to the dismantlement of the Japanese Empire. All possessions outside of the Japanese home islands had to be surrendered and Japan was occupied by US forces. The country was devastated by the war, including by the American bombing campaigns against industrial centres and the use of nuclear weapons against Nagasaki and Hiroshima. But beyond the destruction of the war, Japan had to confront the moral legacy of the conflict. The Japanese military had fought the conflict with exceptional brutality, especially in China. Japan had to contend with a legacy of war crimes and crimes against humanity, whether one considers the Nanking Massacre, the Japanese army's treatment of prisoners of war and civilian population, including scientific experimentation on live test subjects, or the practice of sexual slavery exemplified by comfort women. Japan also suffered the social consequences of mass suicides in order to avoid occupation, underpinned by values associated with the diffusion of *bushido* into society at large.

The military's capturing of the government and the diffusion of a militarist-nationalist ideology in the pre-war period, underpinned by a focus on traditional Japanese social values and loyalty to the Emperor, was identified as a key contributing factor. Post-war Japanese society can be characterized as a strong streak of anti-militarism, underpinned by the trauma of World War II. This is most clearly expressed in Article 9 of the Japanese Constitution that renounces Japan's right to utilize military might as a tool of statecraft. Based on this Japan has been very reluctant to use its military potential even in the context of internationally sanctioned peace-keeping operations or in support of allied forces. Rather Japan has focused on building a foreign policy posture based on economic engagement and regional integration.

This opposition to military power continues until today. Prime Minister Shinzo Abe has sought substantial reforms to the political context of the Japanese Self-Defence Forces (JSDF). At the moment, the JSDF existence is underpinned by a patchwork of loopholes and creative reinterpretations of the constitution. However, putting the appropriateness of the Liberal Democratic Party of Japan's plans, any attempt to alter Japan's defence posture has encountered significant domestic resistance and has been characterized as attempts to drag Japan into another war. World War II has fundamentally altered the sociocultural context of military power and war in Japan.

Changing identities through war

As the above examples show, war has a strong influence on one's identity whether one is the aggressor or the victim of aggression. From a purely policy perspective, a history of war is important because it creates a precedent of past behaviour that can be used as a baseline to anticipate future behaviour. But the importance of war goes beyond that. War can fundamentally restructure identities. A century of aggression turned China from a regional hegemon into a deeply insecure state seeking to recapture a position of regional dominance. In the process it is behaving in a highly threatening manner to other regional states. The legacy of a brutal campaign of aggression in the region has similarly fundamentally changed Japan's identity, shifting from one that internalized militarism during the course of Westernization into a strongly anti-militarist one. Japan is now deeply uncomfortable with the military power it wields.

When studying the rational decision to go to war, it is important to remember that war is a high-impact activity on identity. While war can make sense in a short-term perspective, one should not neglect the recognition that it has long-term consequences. The creation of enemies with deeply entrenched competitive identities is a significant political risk. Japan has sought to leave behind its wartime legacy and to integrate into the international system as a responsible member of the community. But its political ambitions have been frequently hindered by the competitive identities it has created. For example, both Japan and South Korea have strongly opposed Japanese attempts to gain a seat on the UN Security Council. Rebuilding a post-war relationship is a monumental undertaking. It is not impossible. The United States managed to build a cooperative identity with Japan, despite the use of nuclear weapons against China. But a history of aggression especially can create political obstacles long after the utility of the military conflict has ceased. This is not to discount the possibility for the need to resort to war. But rather this should serve as a reminder that war leaves behind deep scars in the international system. Deciding to unleash war should definitely not be reached without sufficient forethought.

Strategic culture

Just as political science has pondered the universal nature of political behaviour, strategic studies have questioned whether the art of warfare is a universal science with concrete laws or whether there are national specific strategic truths. There are universal principles of strategy: The value of the high ground, the effects of weather, or the utility of strategic surprise. These are translatable across a wide range of strategic scenarios. However, nations also build their own distinct strategic cultures. A nation's strategic culture combines specific interpretations of these universal strategic principles as well as idiosyncratic strategic principles unique to the particular country, acquired based on their specific wartime experiences. Strategic cultures can produce fairly unique ways of fighting. This is especially important when discussing war in a rational decision context (see Part IV). Specific strategic cultures can be particularly effective or ineffective against various opponents as the combination of distinct fighting styles creates unique military challenges. One good example is the United States. It is undeniably excellent in joint-operation warfare. If it is allowed to fight in its preferred way, it has dominated the battlefield. However, when the US strategic culture clashed with doctrines emphasizing guerrilla operations or other indirect tactics, the US military often suffered setbacks. During the Vietnam War, the United States had significant difficulties effectively countering the fighting style of the North Vietnamese forces, and by the time it adapted its strategy the popular consensus has turned against the war. Similarly, while the United States has dominated the early phases of the Afghanistan and Iraq wars, quickly destroying the conventional Iraqi army, it struggled in the later stages with insurgency tactics. China, as discussed below, has developed a strategic posture specifically to counter the American way of waging war, which has the potential to create significant strategic conundrums, such as the difficulty of overcoming anti-access defences. The development of distinct strategic cultures stems from both historical experiences and material constraints. The Soviet "new school" of maritime defence – which focused on shore-based assets, submarines, and small coastal combatants – was necessary due to the material weaknesses of the Russian navy to carry out its mission along the lines of traditional naval doctrines.[12]

Analysing strategic culture can not only highlight the distinct ways a country may fight, but also provides insights into their potential for war. The section below will discuss China's "active defence" stance, which is a prime example of this. A country's strategic culture contributes significantly to determine when the use of force is appropriate by establishing thresholds that necessitate a military response. The most basic threshold for a military response is an active attack against the home territory of the country. But this can be affected by strategic culture and its definitions of what constitutes "home territory" or an "attack". One has to consider whether the opponent purely defines these in military terms or whether issues such as economic sanctions are viewed as an attack. The definition of home territory is also an increasingly complicated one as one has to ponder whether areas such as cyberspace count within the definition of home territory. These are important considerations for policy formulation, as when they are input into the decision-making tree, they can increase or decrease the attractiveness of certain policy-making paths. It is important to remember that there is no "one-size-fits-all" solution: The enemy has no obligation to adhere to one's own definitions.

Naturally, it can be difficult to discern a country's strategic culture. There is a strategic advantage to keeping one's warfighting methods obscure, preventing an enemy from devising countermeasures. While some countries are required to be a certain degree of transparent due to their democratic political systems and civilian accountability, this is not true for all. Some of the key strategic conundrum countries (e.g., China or North Korea) are less than forthcoming on defence issues. The analyst cannot expect the strategic culture of a nation to be neatly presented, rather it is often necessary to present one's own interpretation based on research into open-source materials. This is especially so in an academic context when one does not have an intelligence apparatus to rely on. While publications on military doctrine are a useful starting point, one should not underestimate the importance of classical strategic texts in understanding the military behaviour of a country. Unfortunately, these texts are sometimes dismissed as "old", insinuating that due to their anachronistic nature they offer no value for modern analysis. When teaching security studies, students have questioned the relevance of these texts. The question is frequently raised about how texts such as Sun Tzu's *The Art of War* or Clausewitz's *On War* can hold any relevance in contemporary strategic studies, especially as Sun Tzu devotes significant time to discussing issues such as the strategic importance of chariots. Troublingly the relevance of even more modern literature has been questioned. I was once asked about the point of studying Schelling's *Arms and Influence*, since it had been first published in 1966. Culture, including strategic culture, evolves over time. Weapons and organization change, accompanied by changes in tactics and strategy. However, the strategic logic of these classical texts, if divorced from anachronistic historical details, continues to be relevant. Arguing that Sun Tzu does not have an important role in modern strategic thinking, let alone in Chinese strategic thinking specifically, would be a heretical notion. Warfare continuously evolves. Air power, tanks, gunpowder, steel weaponry, precision munitions, or nuclear arms have all shaped warfare radically at one point in time. Yet, strategic considerations such as the importance of the initiative, securing proper terrain for battle, or the force multiplier role of strategic surprise continue to hold the same merit from Sun Tzu to contemporary battlefields.

A frequent theme of this book is that "know thy enemy" is an important maxim for war, even if only considered in an International Relations context. Bounded rationality is an important force in decision making, which necessitates a more constructivist approach to the study of International Relations. Even in an International Relations context it is important to have basic familiarity with a potential opponent's strategic behaviour in order to devise accurate normative models. The remainder of this section will analyse China's strategic culture as a case study to illustrate this point.

Strategy to counter weakness: Chinese strategic culture

The strategic culture of China, as an emerging great power, should hold particular interest to anyone studying international security. How the PRC perceives use of force will have significant implications for the international system as a whole. This section will focus on examining key points in the evolution of Chinese strategic culture to discuss the country's contemporary military doctrine.

Sun Tzu and classical Chinese strategic culture

Any examination of Chinese strategic culture has to start with Sun Tzu, the most prominent of the seven Chinese military classics that are a reading for People's Liberation Army officers. *The Art of War* offers an approach to war that focuses on preparation and the manipulation of the enemy to achieve victory rather than battlefield heroics, a key distinction from some Western strategic cultural concepts. The main thesis of *The Art of War* is that the outcome of battles, and by extension wars, is decided long before combatants meet on the battlefield.

A central idea of Sun Tzu is the relational nature of power. Sun Tzu emphasizes that all aspects of war are only meaningful relative to the opponent. Strength in a vacuum is no indicator of actual might. This thinking will be incorporated into a broad spectrum of later theories. One of Morgenthau's key points in *Politics among Nations* is the similar recognition of the relative nature of power that underpins realist balance of power thinking and security dilemma mechanics. Morgenthau uses the example of the defeat of the French army to illustrate this: While remaining just as powerful as in World War I, where it achieved a victory, it was nevertheless defeated by a German army that gained a relative advantage in power in the intervening years.[13]

As warfare is understood in these relational terms, intelligence and preparation become the focus of strategic planning. A battle is won not by battlefield heroics, but by manoeuvring an opponent into a position of weakness from where victory is impossible to achieve. This necessitates a careful consideration of one's own forces as well as the enemy's plans and military resources. As Sun Tzu argues, knowing the enemy is not a guarantee for victory. Rather one needs to be clear on both on the opponent and one's own capabilities in order to achieve victory. This shifts the emphasis from the combat phase of a military confrontation to the preceding preparation phase. To borrow language from the United States' *Joint Operations Planning Guide*, Sun Tzu's approach to war means that the focus of the conflict is in its Phase 0, which is the pre-conflict phase focusing on creating favourable conditions for war.[14] Ideally, the opponent will recognize the unfavourable conditions of war and cease resistance. A famous maxim of Sun Tzu is that a good general doesn't seek victory after entering battle, but rather he has secured a victory prior to seeking a military engagement.

Essentially, in Sun Tzu's strategic thinking, the manipulation of perceptions takes precedent over a test of mettle. The classical Chinese art of warfare is predominantly psychological in nature. The goal is not to defeat the opponent but to persuade it to acquiesce. The ideal operational posture is to be formless and unknowable, always striking at the opponent at unexpected times where it is the weakest. This is distinct from classical Western strategic thinking that has often emphasized the honour of meeting an opponent on the battlefield. Sun Tzu concerns himself not with such notions of heroism. Successful strategy is based on disrupting the opponent's plans, seeking strategic surprise, concentrating forces to achieve local superiority and striking where and when the opponent expects it least, preferably after dividing its forces by forcing them into a reactionary position. This is achieved through manipulating the opponent's perception of the battlefield in favourable ways. Sun Tzu specifically highlights the importance of deception in order to feign strength where one's forces are the weakest, and to feign weakness where troops are concentrated. Similarly, Sun Tzu argues that to make an opponent move, one should manipulate their perception of one's own forces being near or far away. In a war fought according Sun Tzu's principles, information is always concealed and information dominance is used to manoeuvre and manipulate the opponent.

Unsurprisingly, coercion is a strong component of Sun Tzu's strategic thinking. One can recognize the thinking that will later be formalized in Schelling's *Arms and Influence* as "diplomacy of violence" in *The Art of War*. Sun Tzu argues that one of the effective ways to achieve an objective is to identify what the opponent values and threaten that to force compliance. Again, this strategic thinking emphasizes the importance of a deep and thorough understanding of the opponent, its plans, and its political objectives, in order to turn those against them. Threatening valuable territory is an effective strategy to coerce an opponent to seek negotiations and appeasement, rather than risking the ensuing damage.

Sun Tzu's approach to war works because war is fundamentally envisioned as a political activity, pursued in a limited context. War is an expression of statecraft and Sun Tzu argues against putting the destruction of an opponent as its objective. It should be utilized to secure political objectives or concessions. But the destruction of the opponent is viewed as wasteful. Sun Tzu emphasizes that protracted warfare, including attempting to besiege fortified cities, will lead to the economic exhaustion of the state, which opens it up to attack even if the conflict is won, since it would be significantly weakened by the sacrifice required. Similarly, Sun Tzu seems little reason to seek out the destruction of the opponent's military might if victory can be achieved otherwise. This is a significant deviation from the principles of Clausewitz who argued that the primary method of securing objectives is the destruction of hostile military power.[15] In Sun Tzu one can recognize the elements that are later formalized in various limited war doctrines. The goal is the purposeful application of limited force to achieve defined political objectives.

Geography and positioning are crucial elements in Sun Tzu's strategy. Selecting a favourable battlefield terrain is considered to be a strategic force multiplier. *The Art of War* in general emphasizes the importance of securing and maintaining the initiative, as this allows control over the course of the war. The maintaining of the initiative is needed to prepare strategic surprises. Similarly, being on the defensive is viewed as reactionary, and thus there is an emphasis on being on the offensive to ensure that one can fully exploit conditions. The defender has to guard in all direction. But the one on the offence can concentrate forces and direct it against the weakest point of the defender.

A quote that summarizes well Sun Tzu's approach to strategy is as follows:

> [...] attaining one hundred victories in one hundred battles is not the pinnacle of excellence. Subjugating the enemy's army without fighting is the true pinnacle of excellence.[16]

The classical art of Chinese warfare focuses on achieving goals through the manipulation of the opponent, including through various military means. It emphasizes preparation, deception, and being on the offensive, creating an assertive school of thought. A general following Sun Tzu seeks to be always in control, while pushing the opponent into a reactionary position of confusion from where a defeat is certain.

Mao's People's War

While Mao's communist regime was eager to reject most of China's classical past, including the destruction of China's cultural heritage during the Great

Proletarian Cultural revolution, Mao's strategic thinking has clearly been significantly influenced by classical Chinese strategic thinking, including Sun Tzu. For much of the Cold War, the People's Liberation Army has adopted Mao Zedong's *People's War* doctrine to guide its strategic planning. People's War is primarily a doctrine for total war, responding to the contemporary strategic environment. But it maintains many Chinese characteristics in its execution. The central strategic conundrum of *People's War* has been the material inferiority of China. While the PRC has possessed a large population base, the People's Liberation Army has lagged behind main competitors, such as the United States and the Soviet Union. China also lacked a sufficiently developed industrial base to acquire hardware comparable to that of these countries. The PRC needed to construct a strategic doctrine that could ensure the security of the country despite these weaknesses.

The central component of *People's War* focuses on exploiting China's advantages (terrain and population) to defeat a materially superior opponent. Recognizing that the People's Liberation Army would be outmatched by an invading army, *People's War* makes no attempt at border defence. Instead it proposes a three-stage plan to counter an invading opponent. In the first phase, the People's Liberation Army would draw an invading army into the interior of China. This should not be confused with a retreat. The first phase is dominated by the harassment and strategic delaying of the opponent. The main goal is to exploit modern joint-operation army's dependence on supply lines by extending and scrambling them. The main body of the PLA would draw the enemy into China while engaging in constant counterattacks and harassment. At the same time, militia forces would engage in behind the lines guerrilla activities to severe enemy supply lines and other behind-the-line activities. During this period Chinese forces would rely on their superior knowledge of the terrain and maintaining an offensive spirit to control the conflict. Over time, this is expected to lead to the second phase, where a strategic stalemate is achieved. By this point, due to continuous counter attacks and guerrilla warfare, the opponent is expected to be exhausted and its supply lines scrambled, making further advance impossible. Essentially the enemy offensive has been broken. The third phase focuses on putting the main PLA forces on a full-scale counteroffensive to drive the opponent out of Chinese territory. A weakened and disorganized opponent would have little power to resist, despite its material superiority. This is especially so as the PLA would rely on overwhelming numbers afforded by the large population of the PRC.

People's War carries on Sun Tzu's focus on preparation, the importance of selecting terrain, and maintaining the initiative. One of the principles of Sun Tzu was that, if a battle is inevitable, then one should fight on "death grounds", that is, a battlefield where withdrawal for one's own troops is impossible in order to motivate them to fight to the bitter end. A similar focus on geographical advantage can be observed in *People's War*. By drawing in the opponent into the country, the PLA would seek to maximize its own geographical advantage while depriving it from the opponent. Large-scale militia forces and guerrilla warfare also make any territorial gain made by the opponent dangerous, as none of the territory was safe from subversive activities.

A key point of *People's War* is its focus on the human element. Western strategic thinking often emphasized the superiority of technology over men. In contrast, *People's War* emphasizes the importance of human elements, including political and ideological fervour. During the Korean War, the Chinese relied on infantry tactics, including short shock attacks, to penetrate the defences of materially superior UN forces. Ideological commitment was key to the military culture of the PLA, including an expansive commissariat system throughout all ranks.

The overall execution of this strategic doctrine maintained a focus on securing the initiative, a general offensive spirit and relying on force concentration in order to achieve local superiority. The focus is on exploiting any and all sources of strategic advantage to make up for material weaknesses. The maintenance of the offensive spirit is especially important as it will be replicated in modern Chinese doctrine. Being on the strategic defensive was pursued through adopting an offensive stance on the tactical level. Continuing with Sun Tzu's philosophy, being on the defensive was viewed as reactionary and inherently weaker than seizing the initiative and being on the offensive which affords the selection of the timing and location of battles.

Ultimately, *People's War* carried on Sun Tzu's focus on preparation and careful positioning as the key to victory. To quote Mao himself:

> Without preparedness superiority is not real superiority and there can be no initiative either. Having grasped this point, a force which is inferior but prepared can often defeat a superior enemy by surprise attack.[17]

Kissinger highlights another element to Cold War Chinese strategic culture: An emphasis on the psychological aspects of war. Kissinger argues that China practised a fusion of politics and war, where China employed military power not to win an engagement, but to deal a psychological blow that pushes an opponent to adapt a favourable political shift. Kissinger uses the example of both the Korean and Vietnam wars. In either conflict he argues that China did not seek to win on the battlefield, but to deal a shock to the opponent through a quick sharp military strike that would make the opponent change their political stance.[18] Such an approach would fit well into the framework of traditional Chinese strategic thinking presented by Sun Tzu. War is inherently a continuation of politics. The goal is to achieve a particular political change through any means necessary, and victory on the battlefield is secondary to this. Essentially this is in line with Schelling's principles of the diplomacy of violence: The possibility of destruction is used as a bargaining chip. It puts a much higher emphasis on the coercive side of military power than on its traditional interpretation as a tool to seize and hold objectives bypassing the opponent's will.

Local/limited war under high technology conditions

With the end of the Cold War, the strategic environment of China has changed significantly. With the collapse of the bipolar order the threat of total nuclear war has significantly diminished. China needed to re-evaluate its strategic posture. This was influenced by a few key events. The economic policies of Deng Xiaoping have resulted in a dramatic shift in China's economic and demographic centre of gravity. This meant that the in-depth defence principles of *People's War* could no longer be maintained. At the same time, the 1991 Gulf War demonstrated the destructive potential of modern joint-operation warfare. To respond to the factors China sought to develop a new military doctrine.

The contemporary Chinese doctrine adopts a local/limited war focus. In the post-Cold War environment China anticipates fighting localized small-scale conflicts on the periphery of its own sphere of influence, rather than a total defensive nuclear war. These conflicts would naturally emphasize the achievement of specific political objectives over national survival. Specific attention is given to safeguarding China's maritime interests, including the various territorial disputes on the East and South China Sea. Accordingly, a large portion of the Cold War PLA has been demobilized or transformed into the People's Armed Police. Focus has also shifted from a strong ground forces focus (underpinned by China's material weakness) to afford a more significant role to the navy and air forces. Accordingly, there is an increased focus on the acquisition of indigenous high-tech military hardware, underpinned by China's rapid economic development.

Contemporary doctrine carries on China's focus on maintaining the initiative and tactical offence. One of the key lessons Chinese strategic thinkers drew from the Gulf War is that under current joint-operations conditions once the initiative is lost it becomes extremely difficult to regain. A well-equipped opponent, such as the United States, can easily steamroll a reactive opponent. Current doctrine emphasizes an "active defence" stance. Active defence means that, while China maintains a defensive strategic focus, it will do so offensively on the tactical level.[19] If a military conflict is determined to be inevitable, China will not wait for the opponent to launch the first strike, but rather it will seek to dominate the initial phase of the conflict through offensive military action to overwhelm the opponent and quickly terminate the conflict.[20] This stance carries the principles of Sun Tzu who emphasized the importance of control through maintaining the initiative and the superiority of offence.

Central to the contemporary Chinese doctrine is a deep knowledge of the opponent and designing specific tactics to exploit perceived weaknesses. In the current strategic environment, China identifies the United States as the primary near-tier competitor threat to itself. However, the PLA also recognizes the overall material superiority of the US military. In order to address this, China has adopted weapons and tactics specifically designed to counter the US military. The PLA has devoted significant resources to acquire "trump card" weapon system, such as anti-satellite weapons or anti-ship ballistic missiles.[21] These target specific cornerstones of US strategic doctrine, such as a reliance on rapid satellite communications, smart precision-guided munitions, and aircraft carrier groups. These systems afford deterrence to China by supposedly being designed to effectively counter these key elements of US strategy. When it comes to tactics, China has focused on anti-access and area denial strategies to deny command of the sea and strategic mobility to US forces. In this, China is borrowing from Corbett. The PLA recognizes that they do not have the naval capabilities that could challenge the US Navy for command of the sea. Rather, the focus is not to gain command of the sea but to simply deny it to the United States. This is supported by the fact that in a conflict, China would be in a defensive role with little need to move its own forces across strategic waterways. By devising a system of guided missile destroyers, small coastal combatants, and shore-based missiles and aviation, China can achieve local force concentrations in order to deter the United States from entering littoral waters.[22] The strategy is based on the principles of Sun Tzu, on gaining a deep understanding of both the opponent and one's own forces. Due to the superiority of the US military, China needs to rely on carefully designed asymmetrical military tactics to grasp any strategic advantage they can.

The importance understanding Chinese strategic thinking

It is important to understand how China views war and how it engages in it. The fusion of politics and military power and an assertive offensive-focused military doctrine means that China has a high potential for military escalation. While the PRC has not fought a significant military conflict in the post-Cold War environment, military strength has been frequently used in their foreign policy. There are periodic border clashes with India and China has engaged in military coercion over its maritime territorial claims. There is also significant military preparation for an eventual invasion of Taiwan, especially as continued separation from the mainland remains a prominent political issue on the island. Based on its strategic culture, China has a higher potential to pursue a display of military power as a negotiating tool than one would anticipate in the case of some other countries. In any of China's ongoing disputes, a military confrontation is certainly on the table.

Active defence could be an especially problematic policy conundrum. While there is no significant evidence that China would suffer from a "cult of the offensive" as seen among European powers during World War I, the PLA's continuous readiness to strike first to dominate the early phase of a military conflict means that escalation over disputes presents an additional risk factor. It is unpredictable what would persuade China that a military confrontation is inevitable, triggering an armed response. Fravel emphasizes the role of bargaining power in territorial disputes.[23] In many instances, China has a rather weak position. On the East China Sea, its claims over the Senkaku/Diaoyu Islands are precarious at best with minimal bargaining power. On the South China Sea China has a similarly weak legal claim and its position is underpinned by *de facto* possession, questionable land reclamation, and coercion. Even small events in this context can be perceived as a significant negative shift for China, with the potential to trigger a military response.

If one wants to analyse China's potential for use of force in a critical foreign policy context, it is crucial to be aware of its strategic thinking. These characteristics will be crucial additions to an anticipatory or normative model or in a game theory context. For example, when creating a decision tree, the specifics of Chinese strategic culture mean that one needs to adjust the probability of a military response accordingly. Applying a one-size-fits-all model would result in reduced accuracy. Expecting China, Sweden, and Brazil to behave similarly would be unrealistic. Furthermore, a strategic culture also affects the balance of power. While China is overall materially inferior to the United States, its strategic culture allows it to make up for some of those weaknesses. By adopting a doctrine specifically designed to counter US-styled joint-operations warfare, China can alter the balance of power in its own favour against much of the US-led alliance system, such as Japan, who adopts a similar doctrine or strategic culture.

In the end, Sun Tzu's principle of understanding one's own and the opponent's capabilities continues to ring true. To effectively understand war, even in a foreign policy context, it is important to be familiar with the players. This is especially important in a normative context. Some theories of International Relations and security studies are systemic in focus or universally applicable. But even then, identity-based considerations can produce variations in behaviour that are analytically meaningful, especially in a policy formulation context.

Notes

1 Rawlinson, J. L. (1967) *China's Struggle for Naval Development, 1839–1895*. Harvard University Press: Cambridge.

Swanson, B. (1982) *Eighth Voyage of the Dragon: A History of China's Quest for Seapower*. Naval Institute Press: Annapolis.

2 Callahan, William A. (2010) *China: The Pessoptimist Nation*. Oxford University Press: Oxford.

Wang, Z. (2012) *Never Forget National Humiliation*. Colombia University Press: New York.

3 Fenby, J. (2009) *The Penguin History of Modern China*. Penguin: London.

4 Callahan, William A. (2010) *China: The Pessoptimist Nation*. Oxford University Press: Oxford.

5 Wendt A. (1992) Anarchy Is What States Make of It: The Social Construction of Power Politics. *International Organization* 46(2), pp. 391–425.

6 Chung, C. (2009) "The 'Good Neighbour Policy' in the Context of China's Foreign Relations. *China: An International Journal* 7(1), pp. 107–123.

7 Henshall, K. (2012) *A History of Japan: From Stone Age to Superpower*. Palgrave Macmillan: New York.
8 Ibid.
9 Ibid.
10 Ibid.
11 Ibid.
12 Till, G. (2004) *Seapower: A Guide for the Twenty-First Century*. Frank Cass: London.
13 Morgenthau, H. (2005) *Politics among Nations: The Struggle for Power and Peace*. McGraw-Hill: London.
14 "Joint Operations Planning" by US Joint Chiefs of Staff, 2011. www.dtic.mil/doctrine/new_pubs/jp5_0.pdf (Accessed 11 April 2016).
15 Clausewitz, C. (1874) *On War (Translation by J.J. Graham)*. Guttenberg.
16 Sawyer. R. D. (2007) *The Seven Military Classics of China*. Basic Books: New York, pp. 161.
17 Biao, Lin, ed. (1966) *Quotations from Chairman Mao Tse-Tung*. First ed. Foreign Languages Press: Beijing, pp. 98.
18 Kissinger, H. (2012) *On China*. Penguin: New York.
19 Ministry of National Defence (PRC). (2015) *Defence Policy - III. Strategic Guideline of Active Defense*. Beijing: Ministry of National Defence. http://eng.mod.gov.cn/Database/WhitePapers/2015-05/26/content_4586711.htm (accessed 11 April 2016).
20 Lilley, J. and Shambaugh, D. eds. (1999) *China's Military Faces the Future*. M.E. Sharpe: Armonk.
21 Ng, K. P. (2005) *Interpreting China's Military Power: Doctrine Makes Readiness*. Frank Cass: New York.
22 Cole, B. D. (2010)*The Great Wall at Sea: China's Navy in the Twenty-First Century*, Second ed. Naval Institute Press: Annapolis. See also, Tangredi, S. J. (2013) *Anti-Access Warfare: Countering A2/Ad Strategies.* Naval Institute Press: Annapolis.
23 Fravel, M T. (2007) Power Shifts and Escalation: Explaining China's Use of Force in Territorial Disputes. *International Security* 32(3), pp: 44–83.

4 War, international law, and morality

While war has always been part of the international system, attitudes towards the concept differed greatly at various points in time. Ancient societies like Sparta put martial prowess at the centre of their society. Mediaeval knights romanticized warfare and the culture of chivalry built around it. Japan's samurai code of bushido made honour a central organizing component of society. In the age of absolutism, warfare was mocked as a game for nobles and kings. However, one can observe that as citizenship and mass mobilization became more common, especially with the introduction of conscription and citizenship being tied to military service, attitudes tended to shift away from the nobility of war to focus on the horrors of war and thus the need to regulate warfare. This is especially so as mass mobilization, total war doctrines, and improvements in weapon's technology contributed to increasingly significant damage and death tolls, eventually culminating in global conflicts such as World War I and World War II.

The 1864 Geneva Convention

The earliest examples of concentrated efforts to internationally regulate war have primarily focused on in-war (*in bello*) considerations, rather than attempting to limit the occurrence of war in the system. An early example would be the 1193 decision by the Pope to ban the use of crossbows against fellow Christians due to the increased lethality of the weapon. Many of such early rules concerning warfare were of religious origins. International law, as law, is a fairly recent addition to International Relations. For a long time, the foundations of international law were customary principles followed by European nations, rather than a codified set of rules, sometimes established as part of treaties between specific nations. A major effort to respond to the growing lethality of war was the creation of the First Geneva Convention in 1864. Contemporaries such as Henry Dunant and Florence Nightingale brought the horrors of war in the 19th century into focus, especially when it came to the care of wounded soldiers. While weaponry has developed significantly, increasing lethality and injury, the medical care of soldiers continued to be relatively rudimentary under battlefield conditions. The 1864 Geneva

Convention sought to establish international rules for medical care on the battlefield, importantly creating the rules for (a) rendering medical aid regardless of nationality and (b) creating protections for medical personnel on the battlefield.[1] The convention does not engage with the legality of war itself, rather it mainly seeks to ease its negative consequences by formalizing mutually beneficial rules between combatants. Much of the early customary rules focused on such areas of mutual benefits where needless cruelty or suffering could be avoided.

The Hague Conventions of 1899 and 1907

The Hague Conventions similarly sought to regulate warfare. While the 1864 Geneva Convention was a relatively short agreement of ten articles, the Hague Conventions, titled "Regulations concerning the Laws and Customs of War on Land", are a massive treaty of 60 articles. Some important points of the Hague Conventions are presented below:[2]

- Formalizing the uniformed nature of armies, including distinct uniforms and emblems and carrying arms openly. This is crucial for identifying legitimate targets and remains important until today, especially in the context of insurgents.
- Prisoners of war must be treated humanely and exchanged at the end of the conflict.
- Article 22 states that "the right of belligerents to adopt means of injuring the enemy is not unlimited". The Hague Convention declares certain types of weaponry and tactics as prohibited, including poison weapons, injuring surrendered troops, refusal to accept surrender, the use of weapons exceeding the necessity of force, to use the uniforms of the opponent, to use markings reserved for truce or medical personnel, or to inflict destruction beyond military necessity. The Hague Conventions also prohibit the bombardment of non-defended towns or buildings and require the attacker to issue fair warning. Bombardment in general shall spare places of medical assistance or cultural, religious, or scientific value.
- Pillaging is deemed unlawful.
- Spies are considered separate from soldiers.
- The Conventions create specific rules for armistice and surrender.
- The Hague Conventions also specified rules for occupation of territory, including their administration.

The Hague Conventions embody what is today understood as many of the basic rules of war. Most of these stipulations have been passed down to other conventions, including various Geneva Conventions, and continue to be in effect until today. Once again, the goal was not to regulate the occurrence of war, but to contain the destruction of war within reasonable rules. Much of these rules seek to limit the damage inflicted by war or to codify customs of military honour.

The provisions of the Hague Convention are especially important as military technology has been advancing significantly. Artillery was becoming increasingly destructive at ever-increasing ranges. With an increasing focus on total war approaches, the potential destruction of cities and towns also became a growing reality. The proliferation of machine guns and long-range rifles preceding World War I had significantly increased the lethality of warfare. A good example of the changing times is the 1905 Russo-Japanese War. The Siege of Port Arthur saw significant casualties from the Japanese forces' decision to engage in "human wave" mass infantry attacks against entrenched Russian troops supported by machine guns and artillery.

The Hague Conventions essentially formalized the modern model of warfare: Uniformed troops engaging each other on the battlefield. Soldiers possess the right to surrender and then to be treated as prisoners of war until the end of the conflict. The war is conducted strictly according to the principle of military necessity. Most forms of deceptive behaviour or treachery have been outlawed and excessive damage has been prohibited by the convention. These rules will underpin the approaching global conflict, which nevertheless will violate some of these provisions. World War I will prove to be a conflict that inflicts damage on an unprecedented scale and resorts to truly horrific tactics such as the use of chemical weapons.

The Covenant of the League of Nations

The occurrence of World War I proved that previous attempts to formulate laws of war have been insufficient. Rather than focusing on *in bello* considerations, the attention of international law was shifted to *ad bellum* issues with the intent of attempting to prevent the occurrence of wars. This is a significant departure from previous considerations. Traditionally war was viewed to be the sovereign right of the state. But the establishment of the League of Nations as an international governmental organization sought to establish norms that considered wars to be unacceptable. The treaty states that

> The High Contracting Parties, in order to promote international co-operation and to achieve international peace and security by the acceptance of obligations not to resort to war, by the prescription of open, just and honourable relations between nations, by the firm establishment of the understandings of international law as the actual rule of conduct among Governments, and by the maintenance of justice and a scrupulous respect for all treaty obligations in the dealings of organised peoples with one another, agree to this Covenant of the League of Nations.[3]

In the introductory paragraph the League of Nations establishes both an opposition to war and an acceptance of international law as a formal rule for conduct beyond its customary nature. The main purpose of the League of Nations was to form an international regime capable of deterring the occurrence of war in the system in order to avoid a tragedy similar to that of World War I.

At the heart of the League of Nation was an attempt to normalize peaceful dispute resolution within its members. Article 12 mandated members to seek arbitration or adjudication to their problems and Article 14 has established the Permanent Court of Arbitration as an avenue of dispute resolution. To support peaceful conflict resolution, the Covenant also sought to establish an arms-reduction mechanism in Article 8. Any member of the League of Nations that broke these provisions was to be subject to stringent economic and political sanctions from the entire group. The League has also established a collective defence mechanism: An attack against any member of the group was considered to be an attack against the entirety of the League, with states offering military assistance to each other. The scope of the League was not limited to its members only, but the League of Nations considered the outbreak of all wars to be a concern (Article 11).

While the League of Nations represented common anti-war sentiments after World War I, even in the Covenant it is clear that war was considered to be a component of statecraft. Article 12 concerned itself with peaceful conflict resolution between league members. And while the Covenant sought to establish alternate mechanisms of dispute resolution to reduce the occurrence of war, it does not contain an outright ban on it, despite what the introductory paragraph of the Covenant would suggest. Rather, if a peaceful resolution has been reached, member states agreed not to pursue a military resolution if they disagreed with the outcome for three months (Article 12).

Despite its admirable intentions, the League of Nations proved to be an ineffective organization that fell victim to the political designs of its member states. During its existence it failed to prevent several military conflicts (such as confrontation between China and Japan or the Italian invasion of Abyssinia) and proved to be ineffective in enforcing collective security. The League seemed highly concerned about the potential withdrawal of members and pursued lenient treatment for violations of its core principles to prevent it. And even when a nation, such as Japan, withdrew from the League, no significant punishment was administered. Ultimately, the League proved to be a failure in pacifying the post-World War I international environment, which has seen several military confrontations and the rise of aggressive totalitarian regimes, culminating in the outbreak of World War II. The League of Nations ceased functional operation with the onset of World War II.

The Charter of the United Nations

Following the conclusion of World War II, the League of Nations has been replaced by the United Nations (UN). The UN represents the strongest attempt to eliminate war from the international system. Following the horrors of two world wars and the introduction of nuclear weapons, the Charter of the United Nations is unambivalent about international law's position on war. Article 2 of the UN Charter States that:[4]

> 2.3: All Members shall settle their international disputes by peaceful means in such a manner that international peace and security, and justice, are not endangered.
>
> 2.4: All Members shall refrain in their international relations from the threat or use of force against the territorial integrity or political independence of any state, or in any other manner inconsistent with the Purposes of the United Nations.
>
> 2.6: The Organization shall ensure that states which are not Members of the United Nations act in accordance with these Principles so far as may be necessary for the maintenance of international peace and security.

The UN Charter offers a near-blanket prohibition on the use or threat of use of force for its members, adopting a clearer stance than the League of Nations did. Furthermore, following the League's aspirations to act beyond its membership, the UN considers these rules to be applicable to states even if they are not signatories to the UN Charter if their actions are viewed as a threat to international security. Since in pragmatic terms any war would be an international event that threatened international peace, the UN presents a mandate to enforce peaceful coexistence on a global level. The primary exceptions to these rules are presented in Chapter VII, where the UN reserves the right to engage in the use of force in order to enforce the contents of the Charter:[5]

> 42: Should the Security Council consider that measures provided for in Article 41 [non-military intervention] would be inadequate or have proved to be inadequate, it may take such action by air, sea, or land forces as may be necessary to maintain or restore international peace and security. Such action may include demonstrations, blockade, and other operations by air, sea, or land forces of Members of the United Nations.

The only other exception to this prohibition on the use of force is collective self-defence, as specified by Article 51:[6]

> Nothing in the present Charter shall impair the inherent right of individual or collective self-defence if an armed attack occurs against a Member of the United Nations, until the Security Council has taken measures necessary to maintain international peace and security. Measures taken by Members in the exercise of this right of self-defence shall be immediately reported to the Security Council and shall not in any way affect the authority and responsibility of the Security Council under the present Charter to take at any time such action as it deems necessary in order to maintain or restore international peace and security.

Essentially, under the contemporary international legal regime the use of force is deemed to be an unacceptable way of conflict resolution, unless it was authorized by the UN in order to maintain international peace. Unlike in previous eras when war was a sovereign right of the state, the post-World War II era considers it to be not so. In a community of sovereign equals, conflict resolution through military force is a threat to the stability of the entire system and should be confined to the dustbin of history, according to the UN.

War and international law (conclusion)

Through the sections above one can observe the evolution of international law concerning war from its customary roots to its highly formalized structures under the UN. While originally primarily concerned with *in bello* restrictions, as wars became ever more destructive, a significant interest was built up in regulating *ad bellum* in favour of eliminating war from the system altogether. However, international law is pushing against significant obstacles in the international system. Despite the existence of the UN, the international system remains an anarchic system of sovereign states. Enforcement and the influence of great powers is a crucial problem, as illustrated by the demise of the League of Nations. This is especially so as the UN has never fully realized the enforcement capabilities it once envisioned for itself. The history of the Cold War shows the limits of the UN framework. The Charter was signed and came into effect in 1945. By 1950 North Korea invaded South Korea and soon the United States under the UN authorization intervened in the conflict, leading to a major war.

The prohibition on war failed to be consistently enforced. The United States and the Soviet Union engaged in several proxy wars during the Cold War. Key members of the Security Council have used their veto power to shield themselves or key allies from UN enforcement attempts. The UN's enforcement powers themselves have been limited by their contingency on the willingness of member states to provide troops and funds. As an example of problematic behaviour, the United States not only removed itself from under the jurisdiction of the International Court of Justice, which it was instrumental in founding, but utilized UN resolutions as a cover for its invasion of Iraq in 2003. Despite the attempts, war is hardly gone from the international system.

In 2005 the UN itself approved a more interventionist agenda in the form of the "responsibility to protect" (R2P), which emphasized that the UN should intervene, if necessary militarily, to prevent crimes against humanity:[7]

> In this context, we are prepared to take collective action, in a timely and decisive manner, through the Security Council, in accordance with the Charter, including Chapter VII [the chapter discussing UN protocols for military action], on a case-by-case basis and in cooperation with relevant regional organizations as appropriate, should peaceful means

> be inadequate and national authorities manifestly fail to protect their populations from genocide, war crimes, ethnic cleansing and crimes against humanity.

Essentially, looking at the evolution of international law, one would be more accurate to argue that it sought to restrict the scope of "legitimate" military action, instead of eliminating war altogether. Even the UN recognizes that war holds a unique position in one's arsenal of statecraft, it is often the only tool to subdue an uncooperative opponent. Thus, despite prohibitions, war continues to be a part of the international system. This is especially so as great powers and hegemons seek to subvert international law to legitimize their own policy agendas through their structural power. Ultimately as seen in the case of the League of Nations, and sometimes in the case of the UN, the great powers can bend the rules in their own favour or ignore them without significant consequences. International law is not law in the domestic sense, as there is no higher authority to enforce them with a monopoly on disproportionate violence as seen in a domestic environment. International law is more akin to customary principles dependent on members' willingness to abide by them in the absence of stringent enforcement. This is not to say that international law is worthless. It has proved indispensable in reducing the occurrence of conflicts, especially between minor powers. It establishes important institutions that can facilitate the peaceful resolution of disputes. But it has hardly banished war form the system. Periods of relative peace are just as much, if not more, a result of distributions of power calculations and the presence of nuclear weapons as international law. War is hardly gone from the system and a critical study of it continues to be relevant to contemporary International Relations. This is especially so as peace today is no guarantee for peace tomorrow. Periodic peace within the international system is common. However, this does not mean that war has been eliminated from the system or that it will never return. And when it does, it is best to be prepared.

War and morality

A discussion on the nature of war could not be complete without examining the moral context of war. While in an arguably realist international system issues such as identity or nuclear weapons may seem to have a larger impact on the decision to go to war, it is nevertheless important to examine the moral perspective as the issue of going to war has been often framed from this perspective in discourse. This is especially so as even in a highly competitive environment warmongers do not seem to prosper in the current international system.

It is often argued that states in the international system are amoral and will work with whomever to further their interests. Democracies work with dictatorships, liberal states sponsor religious fanatics, and so on and so forth. Gray argues that:

> Politics, domestic and international, is not a morality tale. Politics is about power: who has it, how to get it and keep it, and what to do with it. It is not about doing good or being right in some ethical sense. States do not often go to war for moral reasons. Statesmen are obliged to protect the vital interests of their community, and those interests do not include the enforcement of justice, or the punishment of evil, both as culturally determined, naturally. The world is awash with injustice. Brutal regimes exist aplenty and they always will.[8]

This is true to a certain extent, but not fully. Even Gray recognizes that morality can play a role in the course of a war. Reputation is important in the system, as this book keeps bringing it up. Insisting on a basic framework of morality can also provide strategic advantages by restricting the choice of an opponent. More amoral as the realist international system gets, it is nevertheless important to understand our existing framework of morality in relation to war, if for no other reason than to manipulate it for one's own advantage.

In discussing the morality of war, just war theory separates considerations into two main categories. *Jus ad bellum* considerations discuss whether combatants engaged in war for just reasons. *Jus in bello* considerations in contrast discuss whether the war was fought according to moral standards. The two are assumed to be logically independent. One can fight an unjust war in accordance with moral standards. And even if one case for fighting is just, it is possible to pursue those goals in an unjust manner.[9] Both of these considerations are relevant for the discussion at hand.

The basic *jus ad bellum* consideration is that the aggression of war is a moral crime. Walzer argues that aggression is a crime because it visits the hell of war onto people.[10] Why war is hell is not hard to understand. It is hellish because of the destruction and death it visits upon people. As such, war is a crime if that carnage is visited upon people without their consent. The moral wrong the aggressor commits is the putting of people into a position where they have to choose between their lives and their rights. War is not a crime if (a) the combatants are willingly engaging in combat or (b) if combatants have the option to withdraw from the battlefield without reprisal. Combatants are assumed to be willingly engaged in combat if they do so motivated by fame or fortune and not by external pressure. Examples of these could be mediaeval knights or renaissance mercenary armies. Modern professional soldiers don't fall into this category for the most part as they are assumed to join out of economic necessity, which is not regarded as a free choice. Outside these examples, an aggressor is guilty of the crime of aggression as breaches of consent by visiting carnage on people who did not consent to it.[11]

A war can only be just if it is fought in opposition to aggression, for example, wars of resistance or collective self-defence. Acts of aggression, regardless of justification, are considered to be immoral. This idea of morality is largely replicated in the UN Convention that prohibits aggression. This interpretation of morality provides a theoretical foundation for the reputation damage discussed earlier. The international community has established norms on the prohibition of use of force as aggression is viewed as a moral crime. Violation of these rules will result in a reputational penalty and the international community collectively acts to deter acts of violence. However, this is simply a moral judgement, not an absolute prohibition. States with sufficient power can nevertheless engage in war and often use their influence to justify their interventions as moral and in the common interest of the international community.

Adopting a realist perspective, from the viewpoint of a hegemon it makes strategic sense to insist on this basic principle. The hegemon(s) largely sets the norms of the international system. Insisting on a prohibition of violence facilitates the maintenance of a basic sense of order. By insisting on violence being unacceptable, a hegemon is engaging in the third face of power: The hegemon shapes other countries' fundamental beliefs, perceptions, and preferences.[12] By exercising this structural power a hegemon can limit the eagerness of lesser powers to engage in various conflicts that could act as a destabilizing influence on the established order. This is especially so if we consider the current global system as the capitalist world economy. The distributed nature of the world economy, capitalism relying on a large geographical scope and a global division of labour, necessitates a certain level of stability and peace within the system.[13] An insistence on the immorality of war offers to the main beneficiaries of the system an opportunity to regulate the behaviour of system at large through their structural influence. Naturally, there will always be outlier states or instances of war, but the frequency is reduced and the international community is more willing to act together to redress the situation. At the same time, hegemons can still engage in use of force, often dressing it up in the language of collective self-defence. A hegemon doesn't go to war to achieve interests. A hegemon goes to war to protect the system and to remove or punish rogue states that would threaten the stability and peace of the international system. Or at least so the hegemon argues. Due to their significant structural influence, hegemonic interests are often conflated with systemic ones.

Moving away from *jus ad bellum* considerations, the most basic principle of *jus in bello* morality is the moral equivalence of combatants: Regardless of which side they fight on, soldiers are considered to be moral equals. The crime of aggression rests with the decision-makers, not with the soldiers on the battlefield. The moral responsibility of soldiers is to fight a war without breaching ethical principles. The two main considerations here are (a) "who is a legitimate target?" and (b) "what are the limits of military engagement if a target is legitimate?" The first is easy to understand. Civilians or those not actively engaged in combat (e.g., the wounded) should not be targeted. This can raise some moral conundrums, such as whether it is acceptable to target sleeping soldiers, but in general it is fairly straightforward. The latter focuses on how much violence one can morally visit upon a legitimate target. The most fundamental test for this is to determine (a) military necessity and (b) proportionality. Violence is considered to be morally justifiable if visiting it upon the enemy is necessary for accomplishing victory and the absolute minimum amount of violence needed to accomplish the objective is used. Additionally, certain tactics and weapons are considered immoral due to their disproportionate brutality. Examples of this are indiscriminate bombardment, cluster bombs, chemical and biological weapons, using flamethrowers on soldiers, torture, and so on and so forth.[14]

Again, from a realist perspective, it makes sense to support such moral prohibitions, including their translation into international law. It constrains the military possibilities of the enemy. Many of the weapons and tactics prohibited are especially gruesome, but they are also cheap and very effective. Insisting on them being an illegitimate weapon of war benefits the major powers of the system. These states tend to have the means to develop substantial military might through "cleaner" means. At the same time, preventing the use of these "illegitimate" weapons and tactics preserves their advantages in the balance of power. One can consider the example of biological and chemical weapons. They are truly terrible instruments of war. They cause indiscriminate damage and often visit a particularly gruesome death on their victims, whether one is talking about mustard gas or weaponized anthrax or Marburg virus. They are weapons of mass destruction. However, their development is relatively low cost, especially with relation to the requirements of a nuclear programme, including the missile programme needed for effective delivery. Thus, they represent an efficient alternative for lesser powers seeking to acquire their own weapons of mass destruction in order to bolster their deterrence. Possessing these weapons would allow them to increase their deterrence against major powers through massively raising the costs of use of force against them. While the prohibition from a humanitarian perspective is obvious, one cannot ignore the power dimensions behind it. Major powers rarely engage in agreements out of the kindness of their hearts. By outlawing these weapons and tactics, they constrain the strategic opportunities of other powers within the system. And if a lesser power acquires such weapons, often in order to mitigate the insecurity arising from being a rogue state within the system, then a morally justifiable *casus belli* is readily available to deal with them.

Overall, one can see that even from a cynical realist perspective it makes sense to insist on a certain basis of morality in the international system. This is not to undermine the importance of condemning the aggression of war as immoral or the need to control the use of certain weapons systems. But it is important to observe that even seemingly innocent elements of the international system are subject to calculations of power. Aristotle argued that political science, as a study of power, is the master science. He did so because all aspects of human existence are subject to power relations, including our sense of morality. Human existence is a socially constructed reality, and its construction is shaped by the distribution of power and influence.

Nuclear weapons

One question frequently pondered is the possibility of war in the shadows of nuclear armaments. The Cold War has been underpinned by the principle of "mutually assured destruction" (MAD). Major powers possess enough nuclear arms not only to defeat any opponent but also to effectively destroy human civilization. The results of total thermonuclear war are expected to be catastrophic. While the possibility of a nuclear winter is a debated subject, the sheer devastation caused by the simultaneous use of humanity's nuclear inventory would be highly destructive. Thus, nuclear weapons are expected to act as a strong deterrent to conventional war. This principle has been incorporated into International Relations theory as well. Mearsheimer's offensive structural realism argues that the only position of true security is hegemony. One of the key characteristics of a hegemon is nuclear supremacy, that is, the possession of a substantial nuclear arsenal in high-survivability second-strike capabilities. The principle of mutually assured destruction means that any war would be ultimately futile. If the opponent is at the brink of losing the conflict, then it would have the option to usher in Armageddon. Various peace theories also highlight the role of nuclear armaments in constraining competition and ushering in a supposedly more peaceful international system post-World War II. The true value of nuclear armaments doesn't rest in their ability to act as agents of final revenge, but lies in their ability to act as a deterrent to war. The Cold War remained "cold" because of the nuclear balance between the United States and the Soviet Union. Under such conditions, is war truly possible in the international system?

Schelling highlights that by the end of World War II and the early years of the Cold War, military planning has been consumed by the idea of "society-destroying" wars: The idea that almost all conflicts are total wars which end when the opponent is thoroughly destroyed and all weapons have been expended.[15] Under such planning, should a war occur between nuclear powers, it would be logical to assume that striking the opponent with the full might of their nuclear arsenal, seeking the immediate and total destruction of the opposing society, would be a likely outcome. This would make war a truly monstrous, world-ending proposition. Schelling highlights that by the early 1960s the United States was proposing an alternative to this strategy which would focus on coercion and deterrence extended into an actual war. Nuclear weapons were not to be used as the opening move of a conflict but rather held in reserve to keep enemy cities hostage, subject to the diplomacy of violence. The potential strategic destruction of their cities is kept as a bargaining chip with the enemy to deter certain military actions and to bring an end to the conflict before the total destruction of societies.[16] Schelling effectively counters the narrative that all wars are necessarily total wars, that there is no reason for an opponent not to throw it all into a conflict. Destruction held in reserve is the motivation to restrain from escalating to total war. This is especially important in the context of nuclear armed combatants, but can apply in various scenarios beyond nuclear armed states.

The use of nuclear weapons, just as all other aspects of international relations, is governed by rational constraints. Or at the very least they should be in a normative context. Nuclear war is the logical endpoint of total war, when the goal is the unconditional surrender or destruction of an opponent. However, not all wars are of a total war scenario. Clausewitz has highlighted this when arguing that the use of force needs to be proportional to the objective.[17]

Even under the conditions of mutually assured destruction, limited conflicts remain a possibility, even among nuclear armed states. Limited wars are conflicts that focus on the accomplishment of a specific political objective and are conducted through use of force short of the total military capacity of the nation. A good example for a limited war scenario is the resolution of territorial disputes, especially in the naval realm. The capturing of disputes islands or reefs can be accomplished with minimal force. During the 1974 Paracel Islands conflict between South Vietnam and China a total of ten naval vessels were deployed. The People's Liberation Army – Navy has relied on small combatants: Four anti-submarine warfare vessels and two mine sweepers. The South Vietnamese Navy has mobilized one escort destroyer, one minesweeper, and two seaplane tenders.[18] These both represent a small fraction of the naval power possessed by each. The conflict was resolved far short of escalating into a total war. During the Korean War, Chinese and North Korean troops, backed by the Soviet Union, have engaged US and UN forces. Despite the involvement of both superpowers and the difficult military situation faced by the United States during the war, nuclear weapons were not used. The conflict remained limited to conventional arms and neither China nor the United States mobilized their total military potential. Military power can also be used in the context of compellence. The United States used force to compel Iraqi forces to withdraw from Kuwait. Again, limited force has been used to accomplish a specific objective.

With the end of the Cold War, total war planning has taken a back seat to managing local and/or limited conflicts. China, for example, has refined its doctrine at the end of the Cold War specifically to address this new emphasis on limited conflicts, which its *People's War* doctrine, focusing on total war, was ill-equipped to manage.[19] The early 20th century focus on the destruction of the enemy as a state is largely gone from modern strategic considerations. This actually makes war more relevant to International Relations. Pursuing short, limited conflicts to resolve conflicts of interests is of far more significance to foreign policy planning than the occurrence of massive military conflicts, such as world wars or hegemonic wars.

One should also consider that nuclear weapons represent only a temporary respite from war. Technological advancements are diminishing the deterrent value of nuclear weapons, intensifying competition between major powers by reigniting tensions frozen by MAD. The United States has been actively developing ballistic missile defences. For example, the Terminal High Altitude Area Defence (THAAD) system seeks to intercept short to medium range ballistic missiles, protecting key military targets and population centres.[20] The key issue is that these systems are designed to be effective against ballistic missiles regardless of payload. While they are effective in countering threats from potential rogue states, such as North Korea, they are equally effective in countering the nuclear second-strike capabilities of major powers such as China or Russia. The deployment of US missile defence systems into the Asia-Pacific and Europe has been interpreted as a threat by both China and Russia as it devalues their second-strike capabilities which are viewed as the cornerstone of their national defence. This has reignited an arms race, with Russia focusing on the development of hypersonic missile systems to avoid interception.[21] While nuclear armaments are a sign of great power status, countries have also been always eager to develop countermeasures to increase their own security. Security underpinned by mutually assured destruction is one revolution in military affairs away from becoming irrelevant. If any country develops a reliable countermeasure to missile systems, then the balance of power would be fundamentally altered until a new class of world-ending weapons can be developed.

Ultimately, while nuclear weapons add an additional dimension to the critical consideration of war, they do not eliminate the occurrence of war from the system. While it is very likely that large-scale conflicts such as World War II would not occur in the current system, such conflicts do not represent the bulk of use of force scenarios. War remains a possibility in the current international system and thus it should be considered in the foreign policy formulation process, within rational confines.

Notes

1 Convention for the Amelioration of the Condition of the Wounded in Armies in the Field. Geneva, 22 August 1864. https://ihl-databases.icrc.org/ihl/INTRO/120 (Accessed 08 June 2020).

2 Convention (II) with Respect to the Laws and Customs of War on Land and its annex: Regulations concerning the Laws and Customs of War on Land. The Hague, 29 July 1899. https://ihl-databases.icrc.org/ihl/INTRO/150 (Accessed 08 June 2020).

Convention (IV) respecting the Laws and Customs of War on Land and its annex: Regulations concerning the Laws and Customs of War on Land. The Hague, 18 October 1907. https://ihl-databases.icrc.org/ihl/INTRO/195

3 The Covenant of the League of Nations (1924) https://avalon.law.yale.edu/20th_century/leagcov.asp (Accessed 08 June 2020).

4 Charter of the United Nations. www.un.org/en/sections/un-charter/chapter-i/index.html (Accessed 08 June 2020).

5 Ibid.

6 Ibid.

7 United Nations. *RESPONSIBILITY TO PROTECT*. www.un.org/en/genocideprevention/about-responsibility-to-protect.shtml (Accessed 08 June 2020).

8 Gray, C. S. (2009) *Fighting Talk: Forty Maxims on War, Peace and Strategy*. Potomac Books, pp. 134–135.
9 Walzer, M. (1977) *Just and Unjust Wars*. Basic Books: New York.
10 Ibid.
11 Ibid., pp. 28
12 Nye, J. S. (2011) *The Future of Power*. Public Affairs: New York.
13 Wallerstein, I. (2006) *World System Analysis: An Introduction*. Duke University Press: Durham.
14 Walzer, M. (1977) *Just and Unjust Wars*. Basic Books: New York.
15 Schelling, T. (2008) *Arms and Influence*. Yale University Press: New Haven, CT.
16 Ibid.
17 Clausewitz, C. (1874) *On War (Translation by J.J. Graham)*. Guttenberg.
18 Toshi Yoshihara. (2016). "The 1974 Paracels Sea Battle". *Naval War College Review*, 69(2), 41–65.
19 Goodwin, P. H.B., "The PLA Faces the Twenty-First Century: Reflections on Technology, Doctrine, Strategy, Operations" in Lilley, James, and David Shambaugh, eds. (1999) *China's Military Faces the Future*. M.E. Sharpe: Armonk. pp. 39–63.
20 Lockheed Martin. *THAAD: Integrated Air and Missile Defense*. www.lockheedmartin.com/en-us/products/thaad.html (Accessed 14 June 2020).
21 BBC (2019) *Russia deploys Avangard hypersonic missile system*. www.bbc.com/news/world-europe-50927648 (Accessed 14 June 2020).

Part I

Conclusion

War and international relations

Part I has explored war in different contexts. It examined the basic conceptual considerations of war, as well as how it fits into a larger legal and sociocultural context. For the purposes of this book, war is defined as organized violence used for a political purpose, in a manner consistent with Schelling's description of direct use of force. It is important to understand war in a holistic sense. War doesn't occur in a vacuum. It is affected by the particular combatants that engage in it and their respective strategic cultures. Similarly, the consequences of war are not limited to the immediate conflict at hand. War will have significant implications for the identity of the combatants, both in how they define their security relationships and how their respective identities evolve.

In an International Relations context, the primary interest in war comes from it being a tool of statecraft. War is a course of action for a state to accomplish specific national interests. And within the foreign policy toolset, war occupies a unique position. War is a tool that can produce political results without compliance from an opponent, bypassing its will. This is distinct from the capabilities of other tools that largely seek to target the opponent's will to bargain for or coerce compliance. Thus, war continues to hold analytical significance in foreign policy consideration. Indeed, even though the United Nations rejects that states should resolve international disputes through the use of force, it has recognized the value of military intervention to accomplish compliance from uncooperative regimes through its own responsibility to protect (R2P) concept. The following chapters adopt a simple premise: If one accepts war to be a tool of statecraft, rather than something distinct from politics, then what are considerations of going to war? Even if one is only interested in maintaining peace, it is important to understand the rational decision to go to war in order to disrupt such calculations. Preventative diplomacy is based on recognizing these calculations and developing effective tools to manipulate them in a way that persuades an opponent to pursue alternate policy paths.

There is a strong moral argument to consider war distinct from politics. War, after all, is terrible. It is always a tragedy when war occurs in the international system. And in popular discourse, such a distinction can be useful in order to prevent the normalization of political violence. However, from a policy analytical perspective, such a distinction would be counterproductive. Distinguishing war from other tools of foreign policy obscures its nature as well as the process by which the decision is made to go to war. Theories of International Relations relating to security are often misinterpreted as endorsements of violence. For example, offensive structural realism is often dismissed as a theoretical framework for aggression, despite the fact that Mearsheimer emphasizes the tragedy of the anarchic nature of the international system, offering no moral endorsement for the occurrence of war. The recognition of the internal mechanics of an anarchic system should not be confused with moral acceptance. But the primary purpose of foreign policy is to ensure the continued existence and security of the state in such an anarchic system. Foreign policy makers' primary responsibility is to devise successful strategies to function in such a system. Keeping normative considerations and how one would like the system to function is important. But it should not compromise one's ability to engage with political reality. That reality is that war, just as crime in a domestic system, is not going away. And just as it is important to understand the criminal mind to maintain domestic order, one needs to delve deep into the rational mind that reached the decision to unleash war in the international system.

Issues for further thinking

This section offers a few questions and scenarios that one can explore to further reflect on the key points made in Part I. For each question, the section provides some key details and points to consider but will ultimately leave the question open-ended.

Does one have the responsibility to protect enemy soldiers?

One should consider a hypothetical scenario: There is a strategically important position occupied by enemy forces that needs to be captured. As a commander one has two options. On the one hand, one can call in a precision strike that will obliterate the enemy position, including the 200 enemy soldiers occupying it. The manoeuvre can be executed without friendly casualties. On the other hand, one can launch a ground attack to capture the position. Calculations show that one is expected to lose around 20 friendly troops, but the overwhelming majority of opposing forces will surrender after the initial attack, allowing their capture. Is it preferable to preserve our forces or to spare the enemy? When examining the morality of war, one has to confront this choice and the answer tends to be divisive.

The common-sense answer tends to emphasize the preserving of friendly forces, rejecting one's responsibility to protect hostiles at the expense of one's own troops. They are the enemy and as such they are assumed to be on the wrong side of the conflict. The enemy is assumed to be morally guilty for the crime of war, while one's own side is perceived as morally righteous. If anything, it is assumed that a commander's duty is first and foremost to his or her own troops. Sacrificing them to save enemy combatants is often viewed as inherently contradictory to the nature of military command. When this question is posed to a group of students, a majority tends to line up on this side.

However, if one reflects on the morality of war through the lens of Walzer, then the answer becomes less straightforward. As discussed in Chapter 5, Walzer distinguishes between *jus ad bellum* and *jus in bello* morality. A consequence of this is that soldiers are considered to be morally equivalent and innocent of the crime of aggression, the responsibility for which lies with their national leaders. Assuming that the soldiers did not commit atrocities during the fight, from a "just war" perspective there is no moral difference between opposing soldiers, even if one of their sides must be unjust. If the enemy is accepted as morally innocent, then it is easier to see the validity of a utilitarian perspective. In such a perspective the goal would be to save the most people possible. If our own and the enemy soldiers are accepted as largely equal, then from this perspective it would make sense to suffer minor losses, however tragic they might be, to avert the larger loss of life.

Morality can be a very difficult subject in the case of war. By nature, mobilization seeks to demonize the opponent in order to push soldiers to the point where they can effectively engage in life or death combat. Under such conditions it is easy to forget the shared humanity of troops on both sides. If anything, as the example of the Christmas Truce shows during World War I, once such a recognition is made it is very difficult to return to the fighting. Thus, commanders may seek to prevent such a recognition to accomplish their strategic objectives. Even in a realist international system, morality is a challenging question of trade-offs: In a reverse scenario we would want our opponent to behave in a moral way and preserve our troops. Thus, even from pure self-interest it makes sense to ponder whether one should maximize short-term efficiency or take a longer-term view in which an insistence on a basic moral code can be beneficial on both sides.

Why Russia is threatened by missile defence systems in Eastern Europe?

The United States has been active in the development of various levels of missile defence systems, parts of which are to be stationed in allied countries for wider coverage. The Russian Federation has reacted very negatively to the moving of these assets into Eastern Europe, for example, to Poland. On the surface it can appear hard to be understood of why Russia would be threatened by a system that is inherently defensive. Defensive structural realism (discussed in the next part) has emphasized the difference between offensive and defensive armaments and argued in favour of the latter to prevent the emergence or worsening of the realist security dilemma. One would assume that a missile defence system cannot be used to attack Russia, only to defend against potential uses of missiles by Russia. However, this ignores the fundamental role of deterrence in international politics.

Russia is not so much concerned that if it wishes to attack Eastern Europe its forces would be less effective, but rather it fears that its ability to deter hostile military action is lowered by the reduced effectiveness of its nuclear deterrent. The ability to engage in diplomacy of violence and to rely on one's ability of inflicting a massive second strike in order to disincentivize an attack is a crucial component of major power politics, especially between superpowers (the United States) and ex-superpowers (Russia). Having a reliable nuclear retaliatory option has underpinned the Cold War-era mutually assured destruction (MAD) system that prevented the Cold War from escalating into an actual "hot" war. Developing a system that would allow one country to reliably counter the nuclear missiles of another would create a condition of practical nuclear hegemony: One country could use nuclear weapons or force in general with relative impunity because the second-strike capability of the opponent would be neutralized.

This is an example of a scenario when even defensive moves in the balance of power politics of the international system can have an adverse effect on overall security. Purposeful vulnerability can be at the centre of international peace as fear is an effective motivator to avoid confrontations. But an invulnerable opponent has nothing to fear. Such a situation also raises the question of a pre-emptive war: It would make sense for Russia to strike (or otherwise sabotage) before the system is completed in order to preserve its nuclear deterrent. There are systems that are not developed on purpose because their potential possession would invite a pre-emptive adverse action. Ultimately, the United States needs to balance between increasing its own security and preserving the security of other states to keep the international system stable.

Is humanitarian intervention justified through force?

We have established from both a moral and a legal perspective the crime of aggression: Disturbing the sovereignty and peace of a state through the offensive application of force is deemed contrary to international norms. But this raises the question of what happens if the sovereign state in question is treating its citizens poorly or engaging in actions contrary to international norms, such as ethnic cleansing, and has proved to be resistant to other forms of persuasion than war. One has to weigh the moral imperative of intervention to prevent further harm against the crime of war.

On the one hand, the case for intervention is clear and is based once again on utilitarian principles of preventing greater harm. As discussed above, this has been developed into the principle of R2P. The international community cannot stand by and simply watch atrocities unfold. We have witnessed the devastating consequences of inaction or lack of sufficiently strong action: The Rwandan Genocide, ethnic cleansing in the former Yugoslavia, or the atrocities against the Rohingya in Myanmar. In each instance the international community failed to adequately respond to prevent a humanitarian catastrophe. The case of Yugoslavia has been especially important in illustrating the problem of lack of will to militarily pacify hostile forces. While peacekeepers were on the ground, often they were unable or unwilling to intervene to prevent crimes against humanity. One can easily understand the frustration: What is the point of maintaining a common military force through the United Nations if it cannot put an end to such atrocities?

On the other hand, the current international system is still largely based on the Westphalian model of state system which is defined by the fundamental sovereignty of the nation state. While this sovereignty has arguably been eroded or diffused in a time of globalization, the principle nevertheless remains fundamental to our understanding of the international system. One can only even conceptualize the diffusion of sovereignty in the context of Westphalian state sovereignty. The fact that states should not interfere with the internal affairs of others, especially through the use of force, is the most fundamental of international norms that underpin the system. It is a system of sovereign equals. No state, regardless of size or influence, shall be entitled to intervene in the affairs of another. This is a question of absolute principles: If we accept any possibility of violation to the rule, then the debate is not about "whether" states should interfere with each other but rather "when" it is acceptable to do so. And that can open a pandora's box of power relations where the dominant will justify their actions even if they are morally unjust. This is especially so as our notions of right and wrong are fundamentally constructed through a sociocultural lens. In light of this, some insist on the inviolability of state sovereignty to prevent domination by the strong, even if the cost is inaction as a response to crisis. For example, the Association of Southeast Asian Nations (ASEAN) has been a staunch supporter of inviolable sovereignty. It is easy to dismiss their stance as a cynical political move to protect their sometimes less than democratic domestic politics. But it is important to understand that ASEAN was created in the first place to prevent the region's domination by any major power and to ensure that local forces are in the driver's seat for regional decision-making. As a collection of minor powers, with a diverse and distinct set of cultural and political values, it is not surprising to see that ASEAN would insist on a respect for sovereignty and not a system where the strong enforce their views.

Once again there is no easy solution to this question. It is very easy to fall into the trap of cognitive dishonesty when one argues for intervention in one case but calls for respect to sovereignty in another, based on personal investment or moral opinion. It is also equally easy to fall into a slippery slope fallacy where limited intervention is unfairly equated with unrestrained violation of sovereignty. But the question is important to consider, especially if one is looking beyond the territorial and other power-related causes of war. While one can argue, often normatively, that international politics is an amoral activity, one should not underestimate the desire of people to do good. However, one should also not forget the old adage: The road to hell is paved by good intentions.

Part II

War doesn't work

The idea that war, that is, the use of military power to accomplish our interests, doesn't work is not a new one. No sooner did somebody emerge to espouse this notion than the first caveman picked up a rock with the intent to take something from the neighbouring group of cavemen. The idea that the use of force to further our interests cannot be an acceptable way of conflict resolution is deeply rooted in our sense of self-preservation and morality. It is a morally repugnant idea that violence, especially large-scale organized violence, could ever be a solution. This is closely related to the fact that war as an activity is terrible. In the song *War,* Edwin Starr famously ponders the utility of war, influenced by the Vietnam War, reaching the conclusion that it serves absolutely no purpose. The idea that war is terrible is an uncontroversial one. This is strongly tied to the idea that something that is horrible cannot be an acceptable way of conducting interstate politics. The fact that such sentiments were enshrined into the charter of the United Nations is hardly surprising (see Chapter 4). The United Nations was formed following a turbulent and violent period of human history that witnessed the outbreak of two world wars and countless examples of just how cruel man can be, for example, the brutality of Japan's conquest of China. The 20th century was far removed from Medieval and Renaissance princes who thought about the nobility of warfare.

This thinking is also present in the literature discussing International Relations theory. Several theories exist that either outright state the futility of war or carry normative implications (or have been interpreted to carry such normative implications) to that effect. This chapter focuses on engaging with these theories: Outlining their key assumptions and arguments concerning the role of military power in the international system and taking a critical look at the merits of these arguments. This is needed because despite humanity's disdain for the barbarity of war, the use of force continues to persist in the international system. The collapse of the Soviet Union and thus the collapse of the Cold War order bolstered anti-war sentiments. Free of the spectre of nuclear annihilation and the binding logic of domino theories, the argument was made that war has no place in the contemporary era of peace. No more will children be needed to be taught to cover under their desks in case of a nuclear strike and an international system of shared norms, trade, and cooperation can be created as the ex-communist countries of Eastern Europe emerge from the ashes of the Warsaw Pact and embrace democracy and market economies. This optimism was quickly shattered, most prominently by the disintegration of Yugoslavia and the resulting ethnic violence and North Atlantic Treaty Organization's (NATO) decision to intervene militarily. In a world of post-Cold War optimism, where many militaries were scaling back and struggling to find a purpose, warplanes were once again flying over the skies of Europe. These planes were sent to enforce the international community's vision for the Balkans, which obviously did not include ethnic cleansing, one ordinance at a time. If we look at the international system today, war and the use of military force are far from gone, despite our efforts to create and enforce a peaceful international order. One has to ponder the rationality of humanity if it keeps relying on a tool with such terrible consequences, especially if that tool is deemed to be ineffective in the first place. While some argue that war simply doesn't work, this chapter offers the notion that, although war doesn't always work, sometimes it does.

As shown in Figure PII.2, the part is divided into three main avenues, which are contained in the three chapters below. In order to critically engage with the arguments that war is obsolete in the contemporary international system, this part reviews theories (the particular works examined discussed below) grouped into three broad themes: The theories arguing in favour of the self-defeating nature of war, the schools of thought that offer alternative foreign policy avenues, and the theoretical approaches of fundamentally reshaping the underlying relationships (and thus mechanics) of the international system. For each theory, the goal of this part is to highlight their arguments critical of war (whether contained in the theory or commonly extrapolated from it) and then demonstrate how they fail to decisively demonstrate the obsolescence of war. For each theory the part highlights that, despite their compelling arguments and valid contributions, war remains a legitimate component of international politics, not a universal solution, but as a tool that can resolve specific but stubborn issues in an anarchic system. It is important to note that this chapter focuses on specific works, as listed in Table PII.1. These works have been selected as they are considered representative of the school of thought at large. Since there are diverse views within a specific school of thought, this makes it easier on the part of the reader to identify concrete tenets and for the chapter to address them in detail.

This fits into the larger argumentative arch of this book. The previous part focused on defining the phenomenon of war as a tool of foreign policy and statecraft. And the next chapter examines the conditions under which war occurs. The argument culminates in a critical analysis of how the rational decision to go to war is made (or how it should be made from a normative perspective). This part plays a key role in providing a foundation to the forthcoming discussion by demonstrating that war is still a relevant discussion to have. In the post-Cold War international system, especially as the major powers are at relative peace, one can have the tendency to view war as having lost relevance for policy formulation. In a world of nuclear weapons and globalization, the idea of interstate war seems archaic, as International Relations scholarship becomes fascinated with non-traditional security threats. War gets delegated to less developed or failed states. This chapter demonstrates that from a theoretical perspective war remains a subject of consideration, even if major hostilities have paused for the time being. This is especially so as tensions once again rise as China seeks to carve out its own sphere of influence, as Russia seeks to reclaim parts of its Cold War relevance, and as cooperation between the United States and European Union falters. Temporary suspension of armed hostilities should not be viewed as the onset of global peace. If for no other reason than because a decrease in vigilance in emphasizing preventative diplomacy can result in the return of hostilities. This chapter should serve as a warning that war is never defeated in the international system, and it is merely contained through continuous conscious effort, based on critical scholarship and effective policy formulation (the subject of this book).

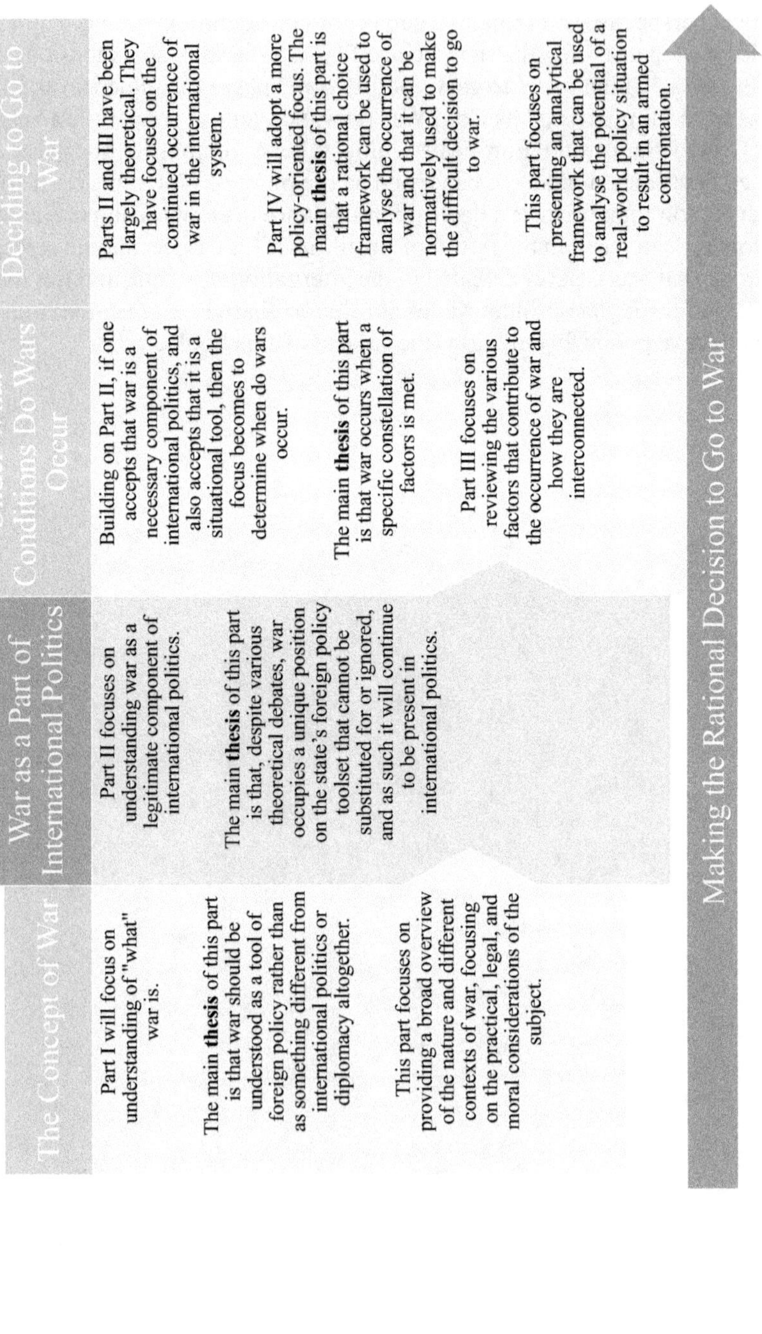

Figure PII.1 Part II's place in the overall arch

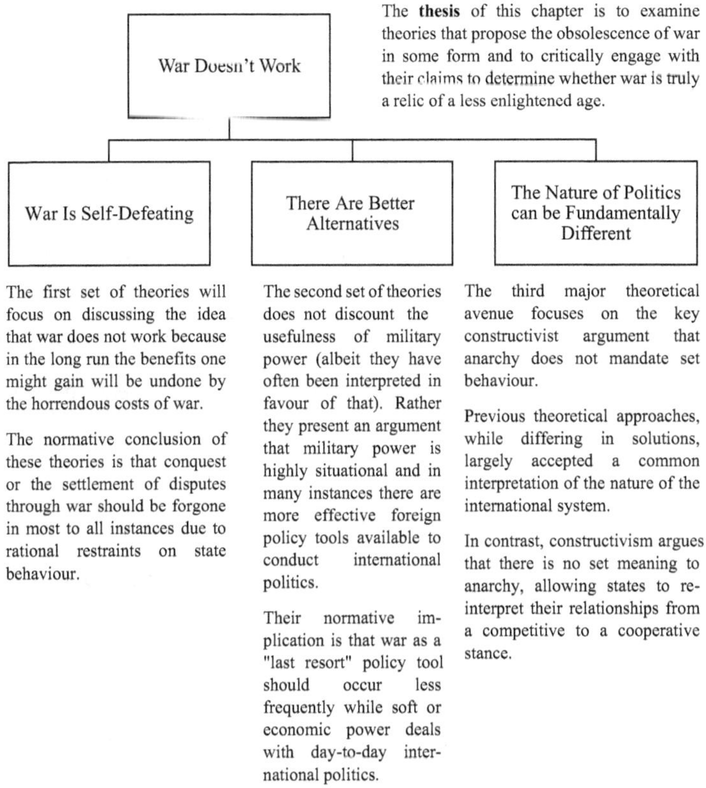

Figure PII.2 Structure of Part II

Table PII.1 Key works for Part II

	Author	*Work*	*Significance*
Chapter 8	Hans J. Morgenthau	Politics among Nations	Foundational text of the realist school of thought after World War II. Morgenthau is the "father" of classical realism.
	Kenneth Waltz	Theories of International Politics	Building on classical realism, Waltz proposes an alternate explanation for state behaviour. This creates the structural realist school of thought.
	Norman Angell	The Great Illusion	While a relatively old book (a product of World War I.) *The Great Illusion* offers an important overview of the roots of economic interdependence as a school of thought, offering the hypothesis that aggressive behaviour can never be sustainable due to the nature of world trade and international economics.
Chapter 9	Robert D. Blackwill and Jennifer M. Harris	War by Other Means	The book provides a critical overview of foreign policy to demonstrate how economic tools can be used instead of hard power to achieve foreign policy objectives. Key text of geo-economic statecraft school of thought gaining relevance in light of the rise of China.
	Joseph S. Nye	Smart Power	A key book exploring "soft power", that is, influence, in opposition to the over-reliance on hard power. Soft power has been a focus of liberal International Relations (IR) scholarship.

Table PII.1 (*cont.*)

	Author	*Work*	*Significance*
Chapter 10	Alexander Wendt	Anarchy Is What States Make of It: The Social Construction of Power Politics	Wendt highlights the predeterministic nature of contemporary IR scholarship, arguing that it's overly restrictive. Social constructivism offers an alternate vision of flexibility and fluidity where interests are constructed rather than extrinsically determined. Social constructivism has gained significant ground in post-Cold War IR scholarship.
	John Mueller	Retreat from Doomsday: The Obsolescence of Major War	Mueller argues that due to the events of the 20th Century and the influence of the peace movement, war has been discarded by the major powers as a tool of conflict resolution.

5 The self-defeating nature of war

As indicated in the Introduction, this part focuses on examining three main theoretical avenues appertaining to the "obsolescence" of war in contemporary International Relations, that is, the idea that there is no space for inter-state war in the contemporary international system. This chapter focuses on reviewing theories that explore the self-defeating nature of war: These theoretical approaches argue (or have been interpreted to argue) that going to war (as the aggressor) never pays in the long run. Hence, states would be better served in engaging in status quo or cooperative behaviour. Two key theoretical approaches are examined here: Defensive structural realism as articulated by Kenneth Waltz (including an examination of classical realism as a foundation) and economic interdependence theory which has roots in the works of Norman Angell. The former is one of the main theories of the realist school of thought, while the latter is a key part of the International Relations liberal tradition. This shows that ideas appertaining to the futility of war are present on both sides of the fundamental International Relations theoretical divide.

Critically understanding these theories is essential for the overall argument of this book. The basic premise of this book is that there is a niche for war even in contemporary International Relations, regardless of the temporary cessation of hostilities between major powers. This is not an endorsement of war as a settlement of international disputes but a call for vigilance for continued preventative policy formulation. That argument is contingent on war being a feasible policy option for states. If it can be theoretically proven that war is a self-defeating activity, then no in-depth discussion on war is necessary: Any state engaging in war as an aggressor is irrational and is led by a regime with a fundamentally flawed cognitive framework, in which case effective policy formulation would be supremely difficult. However, if the chapter can demonstrate through critical engagement the gaps in these approaches that nevertheless allow for the rational occurrence of war within the system, then an in-depth analysis on the occurrence of war is warranted. It is tempting (and morally justifiable) to seek to condemn war to the dustbin of history, but effective policymaking to prevent its reoccurrence is only possible through a critical understanding of the subject.

Classical realism

Realism, emerging after World War II, has traditionally been the domain of competitive power politics and military power. However, realism, despite accusations, was not intended to be a theory of warmongering or a justification for aggressive behaviour. Rather it is a theory of sombre reflection on the seemingly inevitably competitive nature of interstate politics and the continued occurrence of war within the anarchic system. It is a recognition of war, not an endorsement of it. Whether one talks about its classical or structural variants, it is not a theory of conquest, but a theory of survival in a hostile and dangerous international system. While the main focus of this chapter is Waltz's defensive structural realism, this first section will review Morgenthau's classical realism as an introductory point to the larger realist school of thought.

Kenneth Waltz's theory (which is one of the main focuses of this chapter) emerged both as a continuation and a criticism of Hans Morgenthau's classical realism. To understand Waltz's departure, one must first examine the principles presented by Morgenthau and its departure from the liberal International Relations tradition. Morgenthau's classical realism is rooted in a simple idea: It seeks to provide counterpoint to the then prevalent theories of International Relations. Morgenthau viewed these theories' argument that a just and peaceful international order can be realized, and that any failure to do so is a result of individual depravity, ignorance, or greed, to be idealistic. He proposed a contrasting view that conflict is inherent to human interactions, including the international system, and that any attempts to rectify this are doomed to failure. In Morgenthau's view, one should strive to develop policies that work within the confines of the conflictual nature of the international system, policies that allow a state to survive and thrive within the system, not seek to overthrow the system itself. He viewed this approach as "realistic" in opposition to the contemporary idealism of other theories, hence creating the realist school of thought in International Relations.[1] Morgenthau's classical realism has two key components: (1) humans' inherent drive to seek power and dominate and (2) the balance of power governing international relations.

Classical realism, as proposed by Morgenthau, has derived state behaviour from human nature, following a political science tradition reaching back to the Enlightenment. Thomas Hobbes argued that humans are by nature brutish and selfish and require a strong centralized authority for society to function. The anarchy of the state of nature means that fear dominates the thinking of individuals leading to competitive behaviour. Individuals ascend to the social contract to escape the insecurity of anarchy, surrendering some of their sovereignty for security.[2] It is the inherent characteristics of its nature that animate human political behaviour, at the end of the day human beings are little more than political animals. When Morgenthau formulates the foundations of contemporary international realism, he too reaches for human nature as the force that animates political behaviour within the international system.

Morgenthau argues that "the drives to live, to propagate, and to dominate are common to all men".[3] Building on Hobbes' less-than-flattering view of humanity, Morgenthau similarly argues that all men have an innate desire for power – power in this context being defined as the ability to control the behaviour of others – and thus as a political animal they are predisposed to pursue power acquisition as well as to jealously guard whatever power they might have acquired already. According to Morgenthau, these power relations – the desire to gain and exercise power at the expense of others – are present in all human relationships, whether one talks about the workplace, the family, or a social group. Unsurprisingly this drive for power is also translated into the state. Interestingly, Morgenthau calls on man's inability to achieve power to explain the state's drive for power: As most fail to secure substantial power within the domestic structures of the state, they project their desire for domination onto the foreign policy of the state, propelling the state to act aggressively in the international arena.[4]

As the drive for power is inherent to human nature, Morgenthau argues that conflict cannot be escaped in the international system. One may command a wolf to be a lamb and the wolf may even comply for a while, but in the end, it is still a wolf. This is where Morgenthau's international relations depart from Hobbes' domestic politics. Hobbes saw a solution to man's affliction: order through the crushing power of the state. Bringing order to the chaos of anarchy through the surrender of sovereignty. In the Westphalian international system nation states recognize no higher authority, their sovereignty being absolute. But their sovereignty is the root of anarchy: In the international system there is no social contract that would lead to the formation of a hierarchical order. States continue to exist in the state of nature, each being a power onto itself. Morgenthau recognizes the inherent dangers of this state of anarchy.[5] He argues that even if one could "cure" the desire for power within a state, such a state could not survive in the anarchic international system as the other sovereign states, still afflicted by their desire for power, would surely seek to dominate a now defenceless nation. Indeed, unless the desire for power is eliminated simultaneously throughout the entire system, a pragmatically impossible undertaking, it cannot be expected to succeed.

From the perspective of classical realism, war is easy to explain: Interstate relations are always competitive and conflictual. But how does one explain peace in a system of sovereigns, each inherently seeking to destroy the others. In classical realism there is only one thing that limits state behaviour: The amount of power it has relative to opposing states. All states seek to dominate, but not all states possess the power to do so. Morgenthau expresses this in the quintessential realist concept of the balance of power.[6]

What constitutes power and how to measure it appears deceptively simple, yet it remains complex and inaccurate. Morgenthau proposes several components, some of which is relatively straightforward: Geography (the presence of natural boundaries such as mountains or seas), natural resources, industrial capacity, population, and military preparedness. Much of this can be directly quantified (the number of troops, gross domestic product [GDP], etc.) and directly contrasted to determine a balance of power. However, this does not encompass the entire spectrum of power. Morgenthau also identifies more intangible components to power: National character, national morale, and character of diplomacy. As an example, on paper, the United States has enjoyed a significant advantage as a superpower against communist North Vietnam. Yet it failed to translate its economic and military advantage into actual power – that is, the ability to force North Vietnam to concede defeat. North Vietnam's national morale allowed it to emerge victorious from the conflict. The balance of power – who gets to force its will on who – favoured North Vietnam despite its material inferiority. Determining an accurate balance is essential for sovereign states competing for dominance in an anarchic system – later chapters explore this issue further.

States policy calculations – including the decision to acquire military power and to go to war – are dominated by their perceptions of the balance of power. It is important to note that it is always a perceived balance of power. States operate under the condition of imperfect information: At any given time, a state, even one with the surveillance resources of a superpower, has limited information about the power of other states. There are several reasons for this. As mentioned above, some components of power are ill-defined and opaque. However, even when it comes to the more straightforward components of power, the observer encounters significant issues. The military realm is inherently secretive and states dedicate significant resources to keep their capabilities somewhat secret. Sun Tzu has argued that warfare is based on deception and keeping one's capabilities hidden until the time comes to strike.[7] This is partially correct. While modern militaries are secretive about certain capabilities, they also put on bombastic displays of their capabilities as both a deterrent and a coercive force. But these displays, rather than alleviating imperfect information, often compound it, because states are prone to bluffing to exploit the inherent uncertainty of formulating policy based on imperfect information for political or strategic gain. For example, China is highly secretive about its nuclear-armed submarine fleet, actively countering efforts to monitor their activities in the South China Sea. At the same time, China has openly boasted about its unique anti-ship ballistic missile system (ASBM). On the surface, this appears counterproductive and against a literal interpretation of Sun Tzu: By revealing the system, China offers a chance to its rivals to develop countermeasures prior to ever deploying the weapon in combat. To maximize the destructive capabilities of the weapon, it would make sense to seek to keep it as a surprise. But China has revealed the system as psychological warfare: US policymakers being uncertain about the combat effectiveness of the system and the threat it poses to US aircraft carriers are expected to approach the idea of a military confrontation with China more cautiously. The classical dilemma is that, while a secret weapon might inflict more damage, if the enemy underestimates the balance of power, it will be more willing to engage in a war. A war that even if won will prove to be a costly affair. Managing the balance of power under the condition of imperfect information is a careful balancing act: Knowing what to keep hidden, what to share, and what to exaggerate to exploit the weaknesses in the enemy's thinking. The condition of imperfect information is also compounded by the fact that certain vital pieces of information are inherently unknowable. The human mind especially remains an uncharted domain. Intelligence agencies expand significant resources on the profiling of political and military leaders yet claiming to know exactly what Vladimir Putin or Kim Jong Un is thinking or planning at any given moment is largely fantasy.

In the end, there is not even a real balance of power. Or to be more precise, the total balance of power between states is of limited utility. The balance of power is highly psychological: It is a measure of how much a state is willing to bear, not how much it can bear. Once again, we can return to the example of the Vietnam War. The United States could have sustained the conflict. It lost the conflict because it was unwilling to do so. This is exactly why Morgenthau has introduced the component of national morale and national character into the components of power. However, this is complicated by the fact that different states are willing to exert different efforts for different objectives. While the United States was only willing to exert limited effort in South Vietnam, if the communist incursion occurred in Texas, the length the United States would be expected to go would be totally different. The balance of power is absolutely situational and thus the total balance of power is largely meaningless. It assumes the original Clausewitzian notion of total war, from which even Clausewitz backed away from. If we understand war as a political activity, opponents rarely seek the complete destruction or unconditional surrender of their enemies. If balance of power is understood as situational, it is very possible for an overall weaker enemy to nevertheless emerge victorious from a conflict.

With it being arcane to accurately measure and its situational nature, one has to question whether the balance of power as a concept is even worth keeping around. The answer, counterintuitively, is "yes". Foreign policy needs to be formulated, even under the conditions of imperfect information. And while balance of power is not a perfect indicator, it is one that can nevertheless provide valuable information to states about the distribution of power within the international system. While critical engagement with what the balance of power is needed to be for effective application for policy formulation is important, abandoning balance of power as a concept would worsen, not mitigate, the problems of imperfect information. Morgenthau concurs with this assessment. In classical realism, states' thinking is dominated by balance of power calculations. However, Morgenthau recognizes that policymakers are prone to cognitive biases in their interpretation of what the balance of power is. Morgenthau identifies three key biases: First, power is not an absolute but a relative concept. The sentence "Country A has x unit of power" is meaningless. Power is a relative concept and it is only meaningful if it is discussed in terms of having more or less power than an opponent. Second, the distribution of power is not permanent. It is possible that the relative distribution of power changes even if one's own power has not decreased. France emerged as one of the strongest powers in Europe following World War I. However, by World War II, Germany's relative power increased, even though France's total power has not decreased. Third, power is multifaceted and situational and thus attributing undue importance to a single factor is a route to improper policy formulation.[8]

The question of what states seek in a balance of power system is where Morgenthau and Waltz differ significantly. Morgenthau argues that, due to the shifting nature of power, states should not seek an equilibrium, but rather to have a power advantage in order to mitigate miscalculations or sudden shifts in power.[9] As discussed above, the balance of power is more of a perception made under imperfect information, rather than objective reality. One has to anticipate the possibility of miscalculation. And when the stakes are life or death – international politics is played for survival in an anarchic system – Morgenthau counsels that the prudent strategy is to seek a power advantage. This is especially so as even if one has assessed the distribution of power correctly, there is always a possibility for disruption. The introduction of nuclear weapons has fundamentally reshaped the political map following World War II. Similarly, the introduction of computers into the military arsenal led to a "revolution in military affairs" (RMA). The weather wiping out the Spanish Armada has similarly reshaped the balance of power between Spain and other states. Morgenthau's classical realism subscribes to the principle of "better safe than sorry" when it comes to the distribution of power in the international system.

Essentially, a classical realist international system is based on a balance of power where each state seeks to dominate the others by acquiring power. War occurs when there is an imbalance in power: Either the stronger state engages in conquest or weaker states form a coalition to dispatch an emerging stronger state before it becomes too strong. The system is at peace when there is a balance in the distribution of power, but all states seek to overturn it for the sake of their own safety. This is why realist strategic thinking is often consumed by the security dilemma. In its most basic form, the security dilemma is based on the following principles:

1. In an anarchic international system, states seek to acquire power to ensure their security.
2. It is not possible to distinguish between offensive and defensive power acquisition.
3. Observing that their neighbour is seeking power renders other states insecure as they cannot know whether this is done defensively or as a prelude to an attack.
4. Other states begin to seek more power to preserve their security.

Overall, the security dilemma creates a reciprocal dynamic of competition, leaving the entire system more insecure. While each state is individually rational in their pursuit of security, in an anarchic system under the condition of imperfect information this creates a downward spiral of power acquisition and insecurity. This often leads to war if one party cannot keep up.

Defensive structural realism

Kenneth Waltz's structural realism agrees with many of the fundamental assumptions of Morgenthau, but it differs in two key areas: (a) the fundamental force that drives state behaviour and (b) the successful strategies states should adopt within an anarchic international system.

Classical realism's explanation for state behaviour is rooted in Enlightenment-era attempts to extrapolate state behaviour from the individual. We can observe this in both Hobbes and Rousseau, both of whom focused extensively on the "state of nature", that is, the state of humanity before the existence of the state, where man's inherent tendencies were expressed in their purest. Human nature, and its supposed selfishness, is identified at the core competitive behaviour in the international system. War is unavoidable because human nature is ultimately unchangeable. However, the idea of such inherent human behavioural patterns is problematic. It denies the possibility of differentiation between units as well as diminishes the agency of any unit. While Morgenthau recognizes that some of the competitiveness is forced on units through effects such as the security dilemma, at the core of this interpretation of the world is still a desire for power. Waltz breaks with this interpretation. The most fundamental departure of structural realism from classical realism is the recognition that power-seeking behaviour is not a feature or defect of human nature, but rather it is extrinsically motivated, independent from the nature or desires of the agent.

Waltz argues that the internationals system should be recognized and studied as a system[10]. Traditional theories of interstate politics talk about an international system, but it has little meaning beyond a nebulous idea denoting a space within which state-to-state interactions take place. The focus of the analysis has definitely been on unit-to-unit interactions, not the idea of a system in which these interactions take place. Waltz argues that a system should be understood as something that is composed of (a) a structure and (b) units, instead of simply a collection of units.[11] This is an important distinction because Waltz expects the structure of the system to affect the interaction of units as an extrinsic force.

Structural realism if often criticized for reductionism: Eliminating factors that would differentiate units such as culture, national identify, leadership characteristics, etc. While this is a valid concern – understanding the complexities of Sino-Japanese relations is impossible without examining their shared history – it misses the point. Waltz argues that the influence of structural forces can only be revealed if the analysis forgoes unit level differentiation.[12] Structural realism "black-boxes" the state not because state characteristics are unimportant, but because it is an approach designed to reveal forces extrinsic to the state. As such, structural realism is not a stand-alone approach that can answer all questions of international relations. It is a theory that reveals additional considerations ignored by previous analysis. Thus, from a policy formulation perspective it should be used in conjunction with other theories that provide insights into complementary areas of state behaviour. Focusing purely on structural realism would result in an analysis just as inaccurate as an analysis ignoring these systemic forces. Waltz's proposed perspective on the analysis of a system is to focus on "relations" with the system. Waltz recognizes two distinct meanings of "relations": (a) the interaction of units and (b) the relative position of units with the system.[13]

If one adopts Waltz's structural view of the international system, then the question becomes the nature of this structure. The international system is anarchic. This has been recognized by most International Relations theories, including classical realism predating structural realism. Waltz's contribution is not the recognition of the anarchic nature of the international system but formalizing the principles by which this anarchy exerts an extrinsic influence on state behaviour. This was recognized to some degree in earlier works: Classical realism recognizes that anarchy promotes security-seeking behaviour, but it was viewed from the lens of the agent. Waltz turns this around to offer the perspective of the system: Structural realism is less about how one state can survive under anarchy, and more about how the system contends with a collection of states all seeking to survive.

An anarchic system is often hard to comprehend. It has been present in popular culture, but often misrepresented. In popular imagination the idea of anarchy often conjures images of Mad Maxian gangs of leather-bound savages or the isolation of a zombie apocalypse. However, in most of these examples we nevertheless observe the emergence of hierarchical structures familiar to our domestic political understanding. Waltz argues that the nature of a political system can be examined through three key characteristics: (1) The principle by which it is ordered, (2) the specialization and differentiation of units, and (3) the distribution of capabilities across the units.[14] The domestic system is ordered in a hierarchical manner (1). Through the social contract individuals have ceded their sovereignty to the state, which creates rules and governs the state from above the individual. The hierarchical nature of the system allows units to significantly differentiate themselves through the adoption of specializations (2). In the simplest example, some people are farmers, some are soldiers, and some are craftsman. Each has a distinct function within the state and relies on each other. Farmers produce food, but they depend on soldiers for defence and craftsmen for tools. Soldiers provide security, but do not engage in food production or tool manufacturing. And craftsmen create tools to trade for both security and food. This differentiation is allowed by the centralized nature of the state and its hierarchical dominance over the citizens: Farmers can forgo ensuring their own security because the state guarantees it through the creation of a defence force. Domestic order means that citizens can specialize in their contributions to the state. Within a state the distribution of capabilities is relatively equal (3): A citizen represents roughly the same amount of capability as another. It is their specialization that meaningfully distinguishes them, not their relative capabilities.

In contrast, the international system is anarchic (1): States are sovereign, as understood since the Treaty of Westphalia, which means that they recognize no higher authority above themselves and are capable of conducting their affairs without interference from outside. States consider each other equal in their total control of their respective territories. Thus, the international system is decentralized and exists without a governing structure. Waltz argues that "international systems, like economic markets, are individualist in origin, spontaneously generated, and unintended".[15] The state, with its hierarchical order, was purposefully created. There is significant scholarship dedicated to the formation of states and the social contracts that underpin them. Contrary to this, nobody created the international system. It exists on virtue of states existing and it is beyond the conscious control of anyone. It seems self-contradictory that structural realism seeks to divert focus away from state-to-state interactions while also arguing that the international system is called into existence by these state-to-state interactions, with no conscious effort from any force. But Waltz's key point is that the absence of a conscious organizing force nevertheless exerts an influence on state behaviour, exactly through its lack of presence. In an anarchic system states cannot specialize and thus units remain undifferentiated by function (2). The absence of a central organizing power emphasizes self-reliance. The solider and the farmer can trust each other because of the hierarchical power structures of the state. In an anarchic system no such trust exists between states. There is no state that specializes in agriculture while outsourcing its national defence. Similarly, there are no warrior states that create a singular soldier class, forgoing all other functions, and exist as mercenaries for the protection of other states. Since units cannot be differentiated by function, they are distinguished by differentiations in relative capabilities (3). While such a distinction was of little importance in a domestic system, in the international system it is all important. States can be stronger or weaker relative to each other to a significant, and analytically meaningful, degree. Here structural realism reintroduces the realist concept of balance of power. It is the relative amount of power possessed by a state that determines its position within the system, not its specialization as one can observe in the domestic system.

To understand state behaviour, we need to understand both the structure of the system and the motivations of the units. Waltz argues that regardless of culture, ideology, size, or wealth the most fundamental goal of a state is to survive, that is, to continue its sovereign existence. Anarchy incentivizes self-help behaviour due to states not being able to depend on each other. Waltz recognizes that there is always interdependence between states.[16] No state exists within a vacuum and cooperation often provides more efficient means to an end. While development theories of the 1960s and 1970s incentivized complete self-reliance, they also showed the less than pragmatic results such approaches create. It is often prudent for states to cooperate, but realists make the argument that states should not depend too much on such cooperation as (a) in an anarchic system there is no guarantee that a partner will honour deals and (b) cooperation might give power to another state relative to one's own. In matters of security, realism argues for states to ensure their own minimum security through the acquisition of power sufficient to safeguard sovereignty. This leads to the creation of security dilemma dynamics observed by Morgenthau above. However, Waltz' explanation recognizes these dynamics as motivated by systemic forces, that is, the very nature of the international system. Systemic anarchy and the balance of power politics it encourages put an emphasis on competitive behaviour: The acquisition of relative power creates more security, while falling behind renders a state insecure. The result is the aforementioned reciprocal cycle of competition and insecurity. The key contribution of Waltz is to recognize that this is not a conscious process or the result of state decisions, but structural constraints limiting state behaviour. Waltz argues that structural constraints (such as anarchy) limit successful survival strategies, forcing certain behaviour onto states externally.

An immediate solution to present itself: If competitive state behaviour is the result of structural forces emanating from anarchy, then anarchy itself needs to be eliminated to break these forces. If only one would model the international system after the domestic hierarchical order, then the negative cycles of the security dilemma could be broken and an era of cooperation will be ushered. After all, humans have escaped the anarchic state of nature to form their respective states. Waltz is strongly opposed to this view. He argues that all human orders involve some form of violence, regardless of whether they are hierarchical or anarchic.[17] Indeed, the state maintains a law enforcement apparatus through which it exercises its monopoly of force. In a domestic hierarchical society rules are enforced through violence, albeit legalized violence. The police confront lawbreakers with the threat or the use of violence to force compliance. Magnifying this to the international scale, Waltz argues that trying to organize the anarchic international system into a hierarchical one, and then trying to maintain it through enforcement, would require significantly more bloodshed than all the wars fought in an anarchic system combined.[18]

One can derive different normative implications from Waltz's analysis of the influence of structural forces. While under the conditions of structural anarchy states cannot avoid balance of power politics, normatively they should adopt balancing status-quo behaviour instead of seeking to obsessively pursue power advantages. Waltz argues that conquest and expansion are self-defeating under most circumstances.[19] While in the short term one can secure relative power advantages, the long-term costs of this behaviour render it irrational. Reputation is very important in a system governed by imperfect information. In the absence of solid insights, past behaviour is an essential guide in predicting the moves of others. Engaging in expansionist policies will result in a state being labelled as a belligerent actor, amplifying fear and mistrust with other states. Eventually states will create balancing coalitions to counter these perceived rogue states. The system will try to remain in equilibrium, thus the more effort one exerts to gain relative power, the more resistance it will face from the other actors within the system. This is further compounded by Waltz's assertion that there is a diminishing return on investment when it comes to relative power acquisition. Once one has acquired enough power to be in balance, further acquisition is deemed unnecessary. Defensive structural realism argues that being twice as powerful or five times as powerful as an opponent is largely a meaningless distinction, but the difference requires significant investment of resources and creates needless sense of insecurity and reputational penalties with other states. Normatively, states should seek to acquire enough power to safeguard their sovereignty, but they should not seek to overturn the international order. Balance of power mechanics will ensure a relatively peaceful international order. This is a departure from Morgenthau who emphasized power acquisition in excess of other countries to account for potential errors in calculations and potential rapid shifts in the balance of power.

War under this system is often a result of individual irrationality: Some states violate these logical principles and embark on a campaign of aggression to disturb the status quo due to imperfect information or faults in their cognitive framework. Ultimately these states are expected to be defeated and eliminated from the system due to their behaviour contravening structural principles. In Waltz's theory the punishment of aggression is often just as inevitable as the competition for power motivated by structural anarchy. The structure of the international system limits successful behavioural strategies and in Waltz's interpretation the only successful strategy is status-quo behaviour, maintaining the balance of power. Waltz's theory does not propose the absolute elimination of violence. But as a theory it proposes an analytical framework that highlights the futility excessive of power acquisition, and provides a theoretical foundation contradicting the kind of thinking that led to massive interstate wars, including two world wars. Translating defensive structural realism into normative policy formulation, if states recognize the futility of aggression, then a relatively peaceful system can be created. It would be a system of heavily armed sovereigns suspiciously watching each other, but as long as it is understood that structural forces will force a punishment on anyone acting against the status quo, a fragile peace will endure. After all, seeking a war in face of such punishing structural constraints must be irrational, a breakdown of the proper understanding motivated by greed and/or stupidity.

Unfortunately, while the analytical contributions of Waltz are immense, some of the normative promises attributed to defensive structural realism are unlikely to be realized. Defensive realism does not claim to be a theory expressly seeking to end interstate conflict. But based on the principles of the theory, military conflict is normally irrational and is often based on a flawed understanding of the system. In turn, if one accepts that policy should be formulated in a rational manner, then the theory's normative implication is that states should not pursue military conflicts in order to secure interests in most instances (conflicts to counter expansionist states don't violate rational behaviour under defensive realism). This creates a difficult theoretical problem: Offensive wars occur. While Waltz only argues that under defensive realism there are no wide systemic incentives for aggressive acquisition of power, this has been often interpreted as a rejection of war, especially by students. This means that either there are conditions that compromise state rationality or normative interpretations of the theory fail to account for factors when an offensive application of force is rationally justifiable, even under the structural constraints observed by Waltz. If there is a breakdown of rationality, then theory formulation is largely pointless. The irrational by definition is resistant to analysis. One's rationality might be bounded to a cognitive framework unique to the agent (bounded rationality), but as long as it remains rational, it can be understood and ultimately predicted. An irrational agent cannot be effectively understood, and its actions resist analysis and prediction. Instead of engaging with the inherently unknowable, consider a scenario when a rational agent might decide to go against the status-quo behaviour described in defensive structural realism.

The People's Republic of China's economic and demographic composition has shifted significantly following Deng Xiaoping's economic reforms. This had unintended strategic consequences leading to the emergence of interests that prevent China from acting in the status quo manner described above if it is also expected to obey the basic rules of a political realist system. Traditionally, China has been a continental power: Its central provinces were inland and most of its political and security interests presented themselves on land. The primary security considerations for the majority of Chinese history were (a) gaining and maintaining control over the territory that is considered contemporary China and (b) countering external threats emerging primarily from the West (Central Asia) and South (Vietnam and Southeast Asia). The maritime realm was of little significance. Technological limitations combined with the political geography of the region meant that China was relatively safe from maritime threats. The primary maritime threat for the majority of the Chinese Empire has been piracy from the Japanese islands, which posed more of a nuisance than a general threat to the mainland. There were no significant powers occupying the vast seas surrounding China, with the exception of an isolationist Japan. The two main instances where the naval realm was considered significant to China were (a) the Mongol attempts to conquer Japan from China during the Yuan Dynasty and (b) the voyages of Zheng He during the Ming Dynasty. The first instance concluded with the destruction of the Mongol fleet, twice, reinforcing the limitations of large-scale naval warfare at the time. The voyages of Zheng He brought significant riches to China but abruptly ended due to internal political conflicts. The Chinese treasure fleet was burned, putting an end to China's maritime adventures. For all intents and purposes the littoral seas surrounding the Chinese coast constituted a defensive barrier similar to the Great Wall to the west. This continental orientation remained static for most of China's history. While the Middle Kingdom was invaded by European imperial powers from the sea, this proved to be insufficient to shift China's strategic orientation. There was a brief period of naval reform attempts during the final stages of the Qing Dynasty, but they were abandoned in favour of countering continental threats. The 1895 First Sino-Japanese War actively ended China's modern naval ambitions: Japan achieved a decisive naval victory over China and much of the Chinese fleet was destroyed. In 1949 the Communist Party of China (CCP) achieved a decisive victory over the Nationalists, establishing control over mainland China. The military doctrine of the PRC grew out of the experiences of China's War of Resistance against Japan and the CCP's guerrilla struggle against the Nationalists, as well as the political and strategic thinking of Mao Zedong. Mao codified the PRC's military doctrine in *People's War*, which remained the basis of Chinese strategic planning up until the late-1970s. The People's War doctrine put an emphasis on land warfare, with the naval realm being neglected and threated as an auxiliary warfighting domain meant for supporting operations only. The People's Liberation Army – Navy (PLA-N) was meant to conduct harassing operations to slow a potential Western (the United States and allies) advance, but the invasion would have been countered on land by PLA ground forces. In the 1960s and 1970s, the strategic focus of the PLA also shifted from the United States to the Soviet Union following the Sino-Soviet split, further reinforcing the country's continental orientation. In the absence of a significant Chinese maritime presence, other powers have taken strategic ownership of the littoral waters. On the East China Sea, Japan, supported by the US Navy's 7th Fleet, became the dominant force, with Taiwan and South Korea (both US allies) carving out their own spheres of influence. On the South China Sea, the maritime Association of Southeast Asian Nations (ASEAN) countries staked their rival claims, encompassing much of the area. China's maritime presence during this period was mostly limited to small-scale operations conducted against Nationalist troops from Taiwan, clearing their presence from several small islands. China's first major naval conflict was in 1974 when it captured the Paracel Island group from South Vietnam. Although China capitalized on the battle, the conflict was initiated by South Vietnam. China's push into the maritime realm didn't realize in earnest until the late 1970s and 1980s. (The strategic culture and doctrine of China have been discussed in Chapter 3.)

Following the death of Mao, moderates emerged victories from the ensuing leadership struggle. The faction led by Deng Xiaoping sought to end China's revolutionary isolationism in favour of creating a prosperous trading state. Deng's economic reforms have fundamentally shifted the interests and composition of China, thus also altering its strategic outlook. Due to the nature of the reforms, the economic centre of gravity of China has shifted from the hardened central provinces to the traditionally neglected coastal regions. China's new special economic zones sought to feed their growing production into China's commercial ports. This also triggered a wave of internal migration: Rural populations migrated to coastal cities to provide labour to the new industrial sector. This has significantly increased the value of coastal provinces. While they were traditionally considered as expendable buffer zones by Chinese strategic thinkers, they now became the engines of China's economic growth and great power ambitions. This created a new conundrum for Chinese security. The coastal provinces were highly vulnerable and crucial littoral waters were dominated by foreign powers potentially hostile to the rise of China. This was compounded by China's reliance on long sea lines of communications (SLOC) to reach global markets. Essentially, Deng's reforms shifted China from a continental state into a maritime one. The only problem was that China sought a late entry into an already saturated maritime space where major and minor powers have already divided up the space amongst themselves.

This is where China pushes against the logic of status quo behaviour proposed by defensive structural realism. The theory argues that states should seek security, but also that expansion is normally ultimately self-defeating. However, China cannot achieve security without expansion. The current political geography of the region is inherently threatening to China due to its lacking of sufficient defensive buffer between its economically core coastal provinces and other major powers China has identified as a primal threat. The only solution to this conundrum is expansion: China needs to push out its maritime defensive zones and gain control of the littoral waters. However, in a saturated naval space this can only be done at the expense of other states. One can observe this in China's push on the East and South China Seas to establish a solid foothold and to control key waters. Ultimately, to achieve the realist objective of security, China needs to violate the defensive realist theory's supposed normative implications of status quo behaviour. This raises the following question: What happens to states in a defensive realist system that are disadvantaged by the status quo? Their revisionist tendencies are not a product of faulty or irrational thinking but are motivated by the same structural forces that have been identified by the theory. One has to wonder whether such states are simply doomed to a self-defeating cycle where the same forces that motivate their behaviour also create the seeds of their ultimate destruction.

China is not immune to the effects predicted by defensive realism. Its campaign of expansion has attracted significant negative attention, reinforced China threat theories, and triggered balancing behaviour from the United States, Japan, and the ASEAN states. For example, in 1988 China sought to repeat its experience over the Paracel Islands (1974) by encroaching much of the Spratly Islands, but after a short military engagement it was rebuffed by Vietnam. In 1995 China captured the Mischief Reef through military coercion but its actions alarmed regional powers. The 2012 Scarborough Shoal incident pushed the Philippines closer to the US alliance system once again. China's assertive behaviour has triggered the balancing behaviour expected by defensive realism. However, at the same time, one could not argue that China has not benefited from its push into the maritime realm. China has been gaining control over key littoral waters, especially on the South China Sea. And while there are some political costs to its expansion, these costs are not prohibitive. China could successfully leverage its economic importance to soften the political blow from these moves. By exploiting internal divergences of interests in ASEAN, China could prevent a strong unified response. Similarly, Japan's growing security response is now pushing against South Korea, exposing fault lines in the alliance system in Northeast Asia. By pursuing expansion in a gradual manner, never pushing too far in any one instance, China could maintain a balance between gains and costs that would greatly reduce the fatalism of its policies.

Ultimately, while there is often a penalty for expansion, which in the long run can render it self-defeating, it is not inherently fatalistic or irrational. Structural anarchy can and sometimes does require states to expand. And while balancing always exerts a cost, it is not inherently enough to dismiss the use of military force as a foreign policy tool. In a system where the penalty for insecurity is potentially elimination from the system altogether, expansion is sometimes necessitated by the prevailing structural forces, even if there is a significant price to pay. As Waltz argued, existence is a prerequisite to all other considerations. The trick is to keep expansion in balance with the anticipated response to ensure that security-seeking expansion doesn't trigger a stronger threat. The application of military force as a foreign policy tool has to be carefully considered, recognizing the threat posed by balancing behaviour, but it has to be nevertheless considered in an anarchic environment. Interstate wars cannot be simply attributed to "bad apples". It can be a rational response to structural conditions, and as long as it is a rational behaviour it will continue to occur. The normative promise of a status quo system will unfortunately remain unfulfilled as long as the status quo disadvantages some. And in a zero-sum system it will always do so.

Economic interdependence

Economic interdependence as a school of thought encompasses a broad range of ideas and authors. But the central idea of the theory is that the mutual benefit of trade and the resulting interdependent nature of the international system should disincentivize conflicts that disrupt these relationships. This idea reaches far back, as this section's focus on Angell will show, and is rooted in the larger liberal tradition of International Relations. Realism has primarily focused on power politics, military might, and the inherent competitiveness of the system, as shown in the beginning of this section. In contrast, the liberal tradition emphasizes win-win cooperation and the possibility of escaping the competition and violence of the international system through effective policy formulation. Economic interdependence falls firmly within this category. By deepening economic relationships, and thus creating value for both parties, a significant deterrent can be introduced to reduce the temptation to go to war. The argument relies on the assumed rational decision-making of national leaders and their ability to recognize the futility of conflict. While the theory has evolved over time, this section will focus on its roots by engaging with Norman Angell's *The Great Illusion*, an earlier expression of the central ideas of the theory.

Angell, similarly to Waltz, partially emerges as a critic. While Waltz criticized classical realism's focus on human nature instead of systemic force, Angell has been critical of the politics leading up to World War I: The focus on national prestige and interstate competition at the expense of mutual benefits. While Waltz approaches the subject from the perspective of balancing behaviour, Angell argues that whatever limited benefits one might gain from conquest, those are far outweighed by the costs of the endeavour, hence the allure of conquest being nothing more than an illusion. Angell approaches the issue primarily from an economic perspective, rather than the security-dominated perspective of realism. The two key assumptions are that (a) conquest does not actually deliver as much benefits as imagined and (b) the disruption it causes to the international trade system in addition to the material destruction creates a cost that no rational individual could pursue.

In Angell's view conquest is primarily economically motivated: The nation seeks to enrich itself through the acquisition of territory. On this point Angell proposes that such enrichment is hardly possible. If the conqueror seeks to repatriate the wealth from the conquered territory, then it will eventually destroy the value of the territory conquered. And if the wealth is not repatriated to prevent the deterioration of the newly conquered territory, then it remains in the hands of the original inhabitants. Even if one assumes that the state could extract additional tax revenue without disrupting economic activity, the nation is now larger and much of the new revenue would need to be spent on the newly acquired territory. Essentially Angell's argument is that wealth concentrates in the hands of the occupants of a region with limited benefits for the nation overall. Thus, the population changing hands offers little benefit on the national level.[20]

At the same time, war disrupts trade linkages that deliver mutual benefits to both sides.[21] Common in the liberal school of thought, trade is seen as a positive-sum game where both sides benefit. A key component of this is the Smithian logic of natural advantages where certain countries are naturally suited to produce certain goods. Through these natural advantages, efficiency can be maximized, while going against it creates wasteful redundancies. As an example, consider oranges. It is possible to grow oranges in colder climates through green houses. But they require high investment and expensive infrastructure. It is much more efficient to buy them from countries with warmer climates where they can be grown without such investment. Alternatively, one can consider economies of scale: Higher volume production tends to lead to lower per unit price. A system where countries sell what they produce and buy what they need is more efficient than a system where everyone needs to produce everything in small quantities on their own. Specialization and interconnected trade deliver the most efficient economic system, enriching all (at least in theory). Going to war naturally means severing these ties as one does not trade with the enemy. This creates a loss for both sides, beyond the immediate destruction brought on by an armed conflict. Goods become more expensive as the combatants need to replace their trade losses or develop indigenous capacities for them.

Based on this straightforward logic, the theory of economic interdependence has been built up with the core assumption that deepening economic interconnectedness deters conflict. This assumption is based on a rational choice logic: Deepening interconnectedness increases the cost factor of a military confrontation, thus decreasing its overall attractiveness from a cost–benefit perspective, leading to it falling within the rational choice ranking. Essentially the more a state has to lose by seeking a military confrontation, the less likely it will be to do so. This logic has been applied, for example, to the Senkaku/Diaoyu Islands dispute. The dispute is fairly volatile, periodically experiencing strong escalation, especially as China engages in diplomatic sabre rattling. Yet in the past 50 years the dispute remained peaceful. While tensions escalate, they have never reached beyond a certain point before de-escalation takes over. A key argument to explain this trend is that China and Japan maintain a high degree of economic interconnectedness: It would be madness for China to throw away an economic relationship it benefits from for the promise of conquest. The value of the islands, whatever they might be, cannot be compared with the value of mutual economic ties (trade, foreign direct investment, and development assistance). Thus, when escalation reaches a certain point and real damage to mutual ties becomes a pressing concern, China tends to reign in its rhetoric and seek de-escalation.

It is not hard to see the appeal of the theory, especially as it proposes a clearly rational explanation for the futility of war. In this it is similar to defensive structural realism: The argument is not that war cannot happen, but that war should rationally not happen. And it proposes the same analytical problem as discussed above. One has to ask whether some states simply are irrational, and the use of force is a sign of the breakdown of rational choice calculations. Just as in the case of defensive structural realism, this chapter proposes that while economic interdependence can act as a restraining force, it is possible to come to the rational decision to go to war under these conditions.

While wealth is an important consideration for states, it is not the only one, or even the most important one. Economic interdependence applied strictly argues that wealth creation is the primary interest of states, which is a reductionist approach to interstate politics. While wealth is certainly important, both for its own sake and as part of a larger state power, it is arguably not the primary interest of the state. Realism argues that security is the primary interest of the state because the pursuit of any other interest necessitates existence. To put it bluntly, the realist argument is that it is better to be poor but alive than rich but dead – in the interstate context a state that loses its sovereignty or is conquered is considered to be "dead" by Westphalian norms. One has to then consider whether there is a possible scenario when security necessitates expansion through the use of military force, even at the expense of mutual economic ties.

One obvious candidate is the issue of "natural boundaries". Kaplan argues that states have natural boundaries, defined by the geographic maps, that create a secure environment (as discussed in Chapter 9). However, the political map is sometimes out of sync with the geographical map due to various historical conditions, thus pushing states to expand and contract to rectify the disconnect.[22] Various things can constitute natural boundaries, such as shorelines, rivers, and/or mountains. One good example is the case of Russia: Despite its size, much of Russia is geographically open, especially lacking natural barriers on its western border. Furthermore, due to its northern location, it lacks ports that do not freeze over during the winter. The combination of these factors contributes greatly to explaining centuries of Russian expansionism. Russia engaged in conquest, especially in Eastern Europe and Crimea, in search of geographical security deemed essential for the long-term survival of the state. Similarly, China's assertive expansion into the maritime realm can be traced back to demographic changes on the mainland due to Deng's economic reforms shifting the country's economic centre of gravity towards the coast, creating a new strategic conundrum for the defence of the country. In both cases, these countries have pursued expansion through military means even at the expense of economic ties – see Russia's annexation of Crimea and China's grey zone challenges and the 1988 battle at the Spratly Islands. Discounting these scenarios as simply irrational behaviour from belligerent states would be incorrect.

A similar issue is the idea of the possession of resources. Certain resources can be understood as strategic resources, that is, resources that possess importance beyond immediate commercial value. Having access to such resources is a matter of security for the state, and thus conquest might pay dividends even when Angell's considerations of the difficulty of repatriating wealth and of the destruction of trade ties are factored in. One can consider a hypothetical scenario: Uranium. Radioactive materials, such as uranium, are essential for the construction of nuclear armaments, which are in turn the cornerstone of the security of major powers in the international system (even if some indirectly benefit from allied nuclear protection such as Japan). If for a quirk of fate a major power does not possess deposits of its own and the purchase of such materials is not possible through the international markets (either the state cannot compete on price or the sale is blocked by rival powers), then it is a matter of state security to acquire those resources if necessary through conquest, otherwise the state faces a decisive disadvantage in the international system. This logic can be applied to any resource essential for the survival of the state but facing significant scarcity or indivisibility. As environmental degradation became ever more evident, one of the key strategic nightmare scenarios of the future were "water wars", states fighting each other to control the few remaining sources of clean fresh water to ensure that their own citizens have enough. Technology and cooperation can alleviate many of these scarcities – the water wars did not come true so far – but the idea of scarcity is ever present and can provide compelling reasons for expansion, even at the expense of mutual economic ties.

This ultimately leads us to the idea of bounded rationality, a concept this chapter already engaged with above to some extent. Theories such as economic interdependence offer a universally applicable rationality that is useful in a normative context. However, from an explanatory perspective one has to recognize that no such universal rationality exists: Different agents attribute different values to different outcomes. Wealth creation works from the perspective of a secure state. Even Waltz in his insistence that security is the primary interest of the state recognizes that other interests exist once security has been achieved.[23] Above it was highlighted how tangible security considerations can conflict with a focus on economic interdependence. This is just one aspect of a state's insecurity. A state's identity also closely influences what it values and how it relatively values different interests. Economic interdependence is very much rooted in Western political traditions of Smithian liberal economics. One of the key contributions of Alexander Wendt and the social constructivist school of thought is the argument that state interests are not extrinsic, as viewed by both the liberal and realist traditions, but is motivated by the identity of the state.[24] A state's history can contribute to the formation of an identity that is greatly insecure, thus putting a premium on security seeking, just as much as it can create a secure identity that seeks wealth creation within the international system. One good example of this is China. China's identity is based on a mixture of exceptionalism and historical victimhood. The Century of Humiliation narrative is central to contemporary Chinese political identity[25] and is capitalized on by the Chinese Communist Party to facilitate regime survival. This contemporary identity emerged from a history of foreign invasion and territorial insecurity. As such, China is predisposed to view the international system in competitive terms and it views its geopolitical position as fundamentally insecure, more so than the immediate territorial vulnerabilities would indicate. As such, China puts a heavy emphasis on security and is more willing to subordinate its economic tools to political objectives. This partially explains the failure of Japanese foreign policy towards China. One of the key elements of Japanese foreign policy towards China has been economic interdependence to tie China into a web of economic connections and thus reduce the risk of a conflict. However, the effectiveness of this policy proved to be limited. While economic interdependence has been credited with preventing an armed confrontation over the Senkaku/Diaoyu Islands, China has shown a willingness to endure damages to its mutual economic ties with Japan in the service of political objectives, at an increasing rate as its own economic power grows. The normally tightly controlled Chinese public was allowed to engage in popular boycotts and rioting (including damaging Japanese corporate property) over what were seen as Japanese insults to China, resulting in the loss of investor confidence in Japan and the gradual reduction of foreign direct investment (FDI). While so far China has been cautious not to severe economic ties with Japan, it has been increasingly assertive in its foreign policy and security interests, even it meant suffering economic damage. If anything, China has been engaging in coercive politics where damage through economic interdependence is a key bargaining tool, forcing Japan to back down. It is derived from China's willingness to damage mutual economic ties in the service of other, primarily security interests. From the perspective of bounded rationality this is easy to understand. China's political identity is inherently insecure, putting a premium on outcomes that generate power advantages in a zero-sum system. At the same time, Japan's identity is considerably more secure and anti-militarist sentiments put an emphasis on wealth creation in a generally positive-sum system.

This leads to a final point. So far economic interdependence was considered a largely neutral force, with positive-sum results. But this ignores the fact that economic interdependence is subject to its own power dynamics. Above it was already shown that China is keen to exploit its interdependence with Japan in a coercive manner. It can do so because there is a balance of power within the economically interdependent relationship that favours China. While both countries in an interdependent relationship can benefit, if the interdependence is unbalanced, it also allows the stronger to dominate the weaker. If a country is dependent on strategic resources or FDI from another nation, then that relationship compromises the sovereignty of the nation. In such a case, it makes sense for the weaker nation to seek to extract itself from the relationship or the very least alter it to a more favourable state. This can mean military action to secure key interests. Prior to World War II, Japan was dependent on the United States for oil, a key strategic resource. One of the factors that motivated Japan to declare war on the United States was Washington's restriction of trade with Japan due to its aggressive expansion into Asia, which in itself was partially motivated by the resource poor nature of the island nation. It became evident that the United States will attempt to use its advantage in economic interdependence to exert an influence over Japan to which Japan responded through the use of force.

Conclusion

The main takeaway from this chapter should be that, while there are compelling theoretical arguments on why war should be viewed as a highly situational tool that should never be viewed as a first-choice option, these approaches cannot conclusively demonstrate that there is no space for war in contemporary foreign policy. As the examples in this chapter show, there are a number of policy scenarios in which war remains a rational possibility. This connects back to Part 1 of this book: War is a foreign policy tool with a specific purpose, which is to resolve stubborn but important disagreements that have resisted a peaceful resolution for whatever reason. These are usually of vital national importance, for example, related to security or state survival. In these instances, inaction could lead to the elimination of the state itself and as such impermissible for a rational, self-preserving state. While going to war is an option with high costs, it is nevertheless necessary to preserve the state. This does not diminish the consequences of war. A campaign of aggression will likely trigger a balancing behaviour as described by Waltz and it will disrupt mutually beneficial trade ties as proposed by Angell. But this ultimately only illustrates the difficulty of international politics: Choices often have to be made under conditions where all outcomes are negative and the best one can hope is to choose the lesser of two evils. In an anarchic system incentivizing security-seeking behaviour, some damage is sometimes an acceptable trade-off for increased security. If one looks at questions of geopolitics (or geo-security) as examined in Part III or access to vital strategic resources, conquest (aggressive military action) might emerge as a rational solution to the problem, especially if one has exhausted other resolution options or the indivisible nature of the issue makes a compromise impossible. While peace is preferable, a state's primary interest is survival. Any and all other interests are subordinated to this, because at the most basic level the state has to exist to pursue any other interests. Thus, if state survival conflicts with any theoretical principles above, those principles have to yield and war becomes a tangible possibility.

It is important to note that this is not an endorsement of war: War should not be viewed as a desirable outcome to any dispute. But one has to recognize that one cannot rely on the assumption that no state should ever find war as a rational answer to preserve peace and international security. While these theories offer valuable insights into how to affect the key calculations of states, peace nevertheless requires active preventative diplomacy in order to capitalize on these theories and to avoid a scenario in which they would come into direct contact with state survival. Otherwise one will be left with the conundrum of Angell: Knowing that an approaching conflict is irrational and yet being powerless to stop it. The theories discussed in this chapter should be used to support policy efforts to disincentivize conflict but should not be taken as a guarantee that war cannot happen. For example, economic interdependence is a valuable tool to disincentivize conflict between states. But failure to recognize the power dynamics of interdependence and putting absolute faith into its ability to restrain conflicts (as some have done in the context of the Senkaku/Diaoyu Islands dispute) is a recipe for disaster. Trade is important, but wealth is not the only, or even primary, interest of states.

The next chapter moves on from the idea that war is self-defeating and instead focuses on theoretical approaches that seem to offer alternative policy options with more favourable utility, potentially offering a replacement for war. War is terrible, and if we can accomplish the same with a less costly option, then there is no rational reason to rely on war.

Notes

1 Morgenthau, H. (2005) *Politics among Nations: The Struggle for Power and Peace*. McGraw-Hill: London.
2 Hobbes, T (1651) *Leviathan*. Pandora's Box Classics.
3 Morgenthau, H. (2005) *Politics among Nations: The Struggle for Power and Peace*. McGraw-Hill: London.
4 Ibid.
5 Ibid.
6 Ibid.
7 Sun Tzu, *The Art of War* in Sawyer D. (2007) *The Seven Military Classics of Ancient China*. Basic Books: New York.
8 Morgenthau, H. (2005) *Politics among Nations: The Struggle for Power and Peace*. McGraw-Hill: London.
9 Ibid.
10 Waltz, K. (2010) *Theory of International Politics*. Waveland Press: Long Grove.
11 Ibid.
12 Ibid.
13 Ibid.
14 Ibid.
15 Ibid, p. 90.
16 Ibid.
17 Ibid.
18 Ibid.
19 Ibid.
20 Angell, N. (1933) *The Great Illusion*. G.P. Putman's Sons: New York.
21 Ibid.
22 Kaplan, R. D. (2012). *The Revenge of Geography*. Random House: New York.
23 Waltz, K. (2010) *Theory of International Politics*. Waveland Press: Long Grove.
24 Wendt A. (1992) Anarchy Is What States Make of It: The Social Construction of Power Politics. *International Organization* 46(2), pp. 391–425.
25 Callahan, W. (2010) *China the Pessoptimist Nation*. Oxford University Press: Oxford.

6 Better tools available

The second set of theories examined here proposes that there are better ways to secure interests than purely relying on military power. While these theories do not claim to fully replace military power, they have often been interpreted as such, especially by students, ignoring the nuance of the theory. This tendency to oversimplify has been decried by Nye, leading him to coin the term "smart power" to combat the trend of oversimplifying "soft power" and declaring an end to "hard power" based on it.[1] The impulse is understandable. War is terrible and a smart state would not need to rely on such a blunt instrument, or so the argument goes. Influence, rather than brute force, is viewed as the course to take. Essentially, proponents of this approach fall victim to a reverse law of instrumentation. The law of instrumentation, or Maslow's hammer, is a failure in decision-making, arguing that to the man with the hammer everything looks like a nail, that is, that one is likely to seek out familiar tools to solve a problem, even if it is not best suited to it. This attitude is observed by Blackwill and Harris when describing geo-economics: They observe that the US foreign policy has been over-reliant on the use of military force at the expense of other tools of statecraft (in this case, economic influence).[2] Foreign policy formulation encounters a wide range of different situations and indeed should utilize an accordingly diverse toolset to overcome them. This is especially so as countries like China have gained a lot of utility from economic and political influence to pursue key interests without a military confrontation. The debt-trap politics of the Belt and Road Initiative and China's ability to use economic coercion have fascinated observers as Beijing pursues a seemingly bloodless rise to great power status. In international politics not everything is a nail that can be solved with a military hammer. However, it is equally faulty to argue that, just because not everything is nail, nothing is. This section will examine theories of geo-economics and soft/smart power and highlight that, while they coexist with military force within a state's foreign policy arsenal and can offer a better solution to certain problems, a state cannot be purely reliant on influence to function in the international system. On the one hand, military power is needed to back up these tools. On the other, not every problem can be solved through such influence.

Geo-economic statecraft

Geo-economics and economic interdependence from the previous chapter are somewhat interconnected. In an interdependent and increasingly globalized world, where the economy is a strong component of the state's power, they both seek to explore how economic might can be leveraged into political power. However, while interdependence focuses on using connections for mutual benefits and pacifying conflicts, geo-economic statecraft is more interested in economic might as an instrument of power politics. The core concept of "geo-economics" is an argument that overall political objectives can be pursued through the purposeful application of economic instruments, such as trade and investment policies, economic sanctions, and economic assistance. Geo-economics challenges the traditional division of politics and economics that argues that politics seeks power and economics seeks wealth. Instead, geo-economics argues that economic tools can be used to pursue power within the international system. Blackwill and Harris state that:

> Geoeconomics essentially combines the logic of geopolitics with the tools of economics, viewing the economic actions and options of a given state as embedded within larger realities of state power. This fact often puts geoeconomic approaches in tension with assumptions of economics. [3]

The key argument of geo-economics links up with Joseph Nye's arguments concerning smart power (discussed below): States have a vast foreign policy toolbox. However, this diversity of instruments is often taken to a reductionist logical extreme. The emergence or resurgence of foreign policy tools is conflated with arguments about the obsolescence of other ones. Instead of viewing this as the refinement of foreign policy, it is showcased as its reshuffling. This perspective is fundamentally flawed, akin to arguing that the introduction of the screwdriver means the discarding of the hammer. While not everything is a nail, some things are. Relying exclusively on a hammer is equally erroneous to arguing that a hammer is no longer needed. This is especially unfortunate, as Blackwill and Harris present no such argument. While *War by Other Means* highlights the importance of geo-economic tools and advocates for a higher appreciation for them in a primarily US context where the tools are perceived to be underappreciated in favour of a military power-oriented policy, the book does not propose that geo-economics will replace military power or that it cannot coexist with it. Nevertheless, their argument is often misinterpreted in such a manner. If applied to the correct situation, the use of military force can produce results that no other tools in the state's foreign policy arsenal can. But the effectiveness of war is inherently situational. Just as some problems are best suited for a resolution through economic interest, in some cases the state simply needs to fight.

One good case study for geo-economic statecraft is Japan. After the country's defeat in World War II, Japan was demilitarized by the occupying

US forces. Article 9 of the constitution enshrines the country's anti-militarist sentiments:

> Aspiring sincerely to an international peace based on justice and order, the Japanese people forever renounce war as a sovereign right of the nation and the threat or use of force as means of settling international disputes.
>
> In order to accomplish the aim of the preceding paragraph, land, sea, and air forces, as well as other war potential, will never be maintained. The right of belligerency of the state will not be recognized. [4]

Japan is essentially the model of the country that has unilaterally renounced one tool in its foreign policy arsenal. The initial thinking was for Japan to be a fully demilitarized country existing under the protective umbrella of the United States, although this was not fully realized. Post-war Japanese foreign policy has primarily focused on economic instruments: Japan's Official Development Assistance (ODA) programme sought to project Japanese influence through the country's economic strength. This foreign policy stance is based on both liberal and constructivist theoretical assumptions. From the liberal side, economic interdependence theory has been influential on Japanese foreign policy. By creating economic linkages Japan seeks to dissuade conflict by raising the costs of any aggressive action. But the ODA programme has ambitions beyond simply establishing economic interdependence. By increasing contact between countries, Japan seeks to influence how Japan is perceived, shifting neutral or competitive dynamics to a cooperative one, borrowing from the theoretical assumptions of Alexander Wendt.[5] When Japan provided significant assistance to China, the goal was not only to deepen economic links, but to shift the historical narrative between the two countries and to "socialize" a newly emerging PRC into the international system. For example, after the 1989 Tiananmen Square protests and the Communist Party's military intervention, Japan was not in favour of sanctions and lobbied for their speedy lifting due to fears that isolating China would cause it to regress in its integration into the international community. Besides its ODA policy, Japan also puts a high emphasis on foreign direct investment as a tool of influence and possesses considerable soft power. Building on its economic boom in the 1970s and 1980s Japan sought to create a regional order based not on security alliances but production structures: The aim of its "flying geese" paradigm was the integration of Southeast Asian economies into a Japanese-led structure of economic and technological cooperation, which would have afforded significant political influence to Tokyo. Similarly, the Japanese cultural industry has brought influence to Japan: Japanese cultural exports are popular globally and even in societies historically hostile to Japan, such as China or Korea, one can observe significant penetration from Japanese fashion and media. The extent to which Japan relies on non-military instruments to maintain its position as a major regional power is well illustrated by the fact that Tokyo has been often accused of "checkbook diplomacy": Relying on its ability to contribute financially to avoid having to commit hard power instruments to the maintenance of the global order.

One can highlight two main arguments that weaken the argument that geo-economics is a replacement for military power. First, geo-economics is still strongly tied to the military power of a country, especially as a minimum level of military power is needed to maintain state security to engage in geo-economic statecraft. Second, economic influence is highly situational. While a country might possess large economic power, it can only bring geo-economic statecraft to the table when it enjoys a favourable balance of power within the economic interdependence between the two countries.

On the first point, Japan continues to be a good example. Despite being focused on geo-economic statecraft and soft power instruments to exert influence on the global stage, and despite the stipulations of Article 9, Japan is not a demilitarized state. The ink barely had time to dry on the constitution when Japan encountered its first security crisis: In 1950 the Korean War broke out necessitating the withdrawal of US forces that were stationed in Japan to commit to the growing demands of war. It was clear that Japan cannot be left undefended, but it was also equally obvious that the United States cannot alone shoulder the burden of defending Japan. The National Police Reserve force was created in the same year to increase the security of Japan. While officially a police unit, the NPR was modelled after the US infantry and equipped with light infantry weapons. The NPR is essentially the precursor of the modern Japanese military. While Japan is often viewed as a security outlier due to its anti-militarist sentiments, Japan actually maintains one of the strongest military forces in the region. In 1954 the NPR was reorganized into the Japanese Self-Defence Forces (JSDF), re-establishing Japan's hard power instruments. The JSDF represents significant military power: The Ground Self-Defence Forces (GSDF) maintains a 150,000 strong modern force integrated with the US forces stationed in the country. More significantly, the Maritime Self-Defence Forces (MSDF) represents one of the strongest naval forces in the Asia-Pacific with a large power advantage over the People's Liberation Army Navy (PLA-N). The fleet maintains advanced guided missile destroyers based on a combination of indigenous and US technology and recently commissioned carriers of its own, capable of operating helicopters and the F-35B Lighting II combat aircraft. As a regional actor Japan possesses sufficient hard power to engage and potentially win a military confrontation with any other regional power, even without significant US assistance. On paper the military is not a military to abide by the constitution. But in reality, despite its focus on alternate foreign policy instruments, Japan has been working on amassing sufficient military power to ensure the security of the country.

This argument goes back to the basics of realist political thinking. Waltz argued that states have a diverse set of interests. However, their most fundamental interest is security. Survival in an anarchic international system is needed to pursue any other interest.[6] In an anarchic international system, under the conditions of imperfect information, "rich" states, those who could best execute a geo-strategic policy, would present key targets for a military offensive. This is highlighted by Waltz when discussing that structural anarchy prevents the functional differentiation of states: It is not possible for a state to specialize in only one aspect of power.[7] In an anarchic system, self-help represents the only source of security, which means that a state has to take care of its own security first and foremost. To successfully engage in geo-economic statecraft, they need to establish sufficient military power to deter offensive military action. Such a baseline of security is necessarily for the state to be able to purse sovereign policy direction. In an anarchic system, wealth without security cannot be sustained. Classical arguments for the social contract specifically identify this conundrum as why individuals would be willing to give up part of their sovereignty in exchange for security from the state.

The issue of security is compounded by the fact that geo-economic statecraft is often coercive in nature, prompting the target state to seek to escape the effects of the policy. Geo-economic statecraft employs a variety of tools:[8]

- Trade policy
- Investment policy
- Economic sanctions
- Cyber tools
- Economic assistance programmes
- Financial and monetary policy
- National policies concerning natural resources.

Each of these policy options represents a different level of cooperation or coercion to shape foreign policy outcomes. Sanctions are obviously the most coercive one: They seek to directly influence foreign policy through economic damage, based on the principles of Schelling's diplomacy of violence.[9] The target state will naturally seek to circumvent the sanctions, as will several other states either politically invested in the situation or seeking to profit. The enforcement of sanctions has to be backed up by hard power to achieve the desired policy change. While sanctions are often focusing on preventing the flow of goods from one's own country to the target nation, more serious forms of economic sanctions seek to blockade the target nation by controlling the in-flow of specific goods (see the economic sanctions against North Korea) which is highly dependent on coercive enforcement through hard power (not purely military units such as the Coast Guard still constitute hard power and are often backed by actual military might).

Leading to the second limitation of geo-economics, the coercive power of the above-mentioned policy options can be limited by certain factors. The overall effectiveness of geo-economics is affected by the balance of power within the interdependent relationship. As discussed above, economic interdependence is not a neutral relationship, but there is a balance of power encoded within it. In order to coercively employ geo-economic statecraft strategies, one has to enjoy a sufficient advantage in this balance of power, just as one would need a favourable balance of military power in order to pursue a policy of use of force. Coercion inherently implies that one can afford to hurt the other state, and suffer the resulting damage to economic interdependence, and come out ahead of the opponent. If the balance of economic interdependence is unfavourable, geo-economic statecraft offers limited ability to alter foreign policy in a manner an opponent is unwilling to do on their own (building on Morgenthau's basic definition of power that it is the ability to make someone do something they don't want to[10]). This is further compounded by the fact that the balance of power is highly situational: It is only meaningful in the context of specific economic relationships. A country can be overall economically strong and still face limits in its ability to exert economic influence if economic interdependence is limited between the two countries. The coercive ability of geo-economic statecraft is further affected by the target regime's resistance to coercion in general and the degree of resistance to the specific policy change requested. The economic damage potentially inflicted has to outweigh what is asked in return. And importantly this calculation is bounded to the rationality of the target state, so even if it is one's own perspective that the threat is sufficient, it might not be so from the target's perspective. Overall, while geo-economic statecraft can present a route to accomplish certain policy objectives, just as use of force, it is not a solution to all issues.

A good example of geo-economic statecraft succeeding is the 2010 fishing captain incident between China and Japan. In 2010 a Chinese fishing vessel entered Japanese waters and was confronted by a Japanese Coast Guard (JCG) vessel. The captain rammed the coast guard vessel, upon which he was arrested by the Japanese authorities. China demanded the release of the captain on the grounds that Japan has no authority to conduct law enforcement in Chinese-claimed waters. Japan initially refused to do so. As tensions escalated, China was rumoured to be considering a prohibition on selling rare earth minerals to Japan, weaponizing their economic interdependence. Japan's high technology industries are highly dependent on these rare earth minerals and at that point in time China had a virtual monopoly on them. Even if alternate sources would have been available, it would have disrupted Japanese supply chains, causing significant damage. Tokyo decided to repatriate the captain to China. In this scenario we can see that (a) China enjoyed a sufficiently positive balance of economic interdependence and (b) the resistance of Japan was relatively low. Based on these factors the coercive application of economic influence was successful. The release of the captain, while an international humiliation, was a minor issue compared to the real damage a Chinese intervention into the trade relationship could have caused. And Tokyo has been traditionally hesitant to take a hard-line approach against China, motivated by both the strong economic interdependence between the two countries (including significant FDI and outsourcing to China) and

Japan's constructivist foreign policy approach. In this case, geo-economic statecraft through trade policy was able to achieve a policy change.

In contrast, an example that highlights the limits of geo-economic statecraft is North Korea. The United States has put pressure on North Korea to force Pyongyang to abandon its nuclear ambitions for an extended period of time with little success. Through the UN sweeping economic sanctions have been put in place to coerce compliance. However, these were hindered by a number of factors. The United States has limited economic interdependence with North Korea, meaning that it can only exert limited direct influence on the regime. At the same time, the regime with the highest direct economic interdependence, China, has shown limited willingness to enforce these sanctions as they are often contrary to their own political objectives (preventing a regime collapse in North Korea and avoiding a refugee crisis). This is compounded by the fact that North Korea has developed a significant capability for smuggling, circumventing some of the sanctions. The challenge of coercion is further increased by the fact that the regime shows little concern over the suffering caused by these sanctions to ordinary citizens and avoids much of the higher-level inconvenience through political manoeuvring and smuggling. Unlike the scenario above, where high economic interdependence and low regime resistance made Japan an ideal target for geo-economic statecraft, low economic interdependence and high regime resistance allow North Korea to be highly defiant of geo-economic efforts. If the United States wants to achieve North Korea's denuclearization, it seems unlikely that anything short of use of force would achieve that outcome.

The key point here is that, just like military power, geo-economic strategies are situational and are by no means universally applicable. While under the right conditions they can be highly effective, they offer no direct replacement for military power and use of force in the foreign policy arsenal. While Blackwill and Harris make an important point concerning the US over-reliance on use of force, especially in the context of two unsuccessful and costly wars, the argument should not be taken to the logical extreme by arguing that geo-economic statecraft could render military power obsolete.

Smart power

Similarly to geo-economic statecraft, Nye argues that power is more complex and can be exercised in more ways than realism would suggest.[11] Theories such as offensive structural realism focus on largely a single facet of power: Hard or military power. The balance of power is understood to be the primary determinant of what a state can do. In a sense, realist analysis often restricts itself to view power narrowly what Schelling recognized as military force's ability to "seize and hold" interests.[12] This is the most traditional understanding of power, to get something by completely ignoring the opponent's will. This view has been challenged for a long time. Schelling, for example, identified how war can be used coercively, engaging in what he termed as diplomacy of violence, where the goal is not the capturing of objectives but the psychological dominance of the opponent through destruction and suffering.[13] Nye similarly argues that power has to be understood in a more complex manner. He recognizes the basic idea of power: The ability to get somebody to do something we want. But he argues that in execution one has a diverse range of opportunities to exercise such power.

Nye argues that power is a relative and contextual concept.[14] One cannot talk about a state having x amount of power that can be universally translated into all situations, as unnuanced interpretations of the balance of power would suggest. Power inherently depends on the target and what one seeks to achieve. Unless used in its bluntest form to seize things, the use of power is bargaining, without universal applicability. As Nye argues:

> Any attempt to develop a single index of power is doomed to fail because power depends upon human relationships that wary in different contexts. Whereas money can be used to measure purchasing power across different markers, there is no standard of value that can summarize all relationships and contexts to produce an agreed overall power total.[15]

Power has to be understood in a particular context. One has to consider the question of "one has the power to do what?" because traditional material definitions of power – that focus on territory, wealth, or size of military – often fail to take power conversion into account. I encountered this problem when preparing my book on the Senkaku/Diaoyu Islands dispute and China's foreign policy options. One could easily measure the total balance of power between China and Japan through these traditional concepts. The Chinese People's Liberation Army Navy had 15 destroyers while the Japanese MSDF had 51. But this information was largely analytically useless. On the surface this would have indicated that Japan was almost four times stronger than China when it came to naval power – and this was already a more nuanced application of balance of power than simply looking at total military capacity. But this assumed that the conflict would be a total war with the mobilization of all the military assets of a nation, which was obviously not the case. Corbett specifically argued about the suitability of the naval realm to wage limited wars.[16] In a limited war context, total military strength was irrelevant. The question was how much Japan and China are willing to mobilize and commit to the conflict. Once such elements were taken into account, the balance of power appeared more favourable for China than pure material factors would have suggested. This is why Nye argues that taking the material elements of power and translating them into political gains, the actual objective of use of power, is not straightforward. This issue is also identified by Gray when he writes about the importance of good strategy to translate between high-level political objectives and on-the-ground military tactical execution.[17] While Gray calls this strategy, Nye calls the designing of policies with the awareness of complex power conversion "smart power".

Smart power is a combination of hard and soft power assets across multiple relational dimensions to achieve best effect. Nye describes the "three faces" of power as commanding change, controlling agendas and establishing preferences.[18] The first one is the most straightforward, as it largely can be described as compellence. It most closely overlaps with traditional understanding of power, employing coercive influence to make someone do something through the strategic applications of rewards and punishments. In contrast, the second and third faces of power are more indirect. The second face focuses on controlling the agenda of international action in a way that limits the opponent's strategic options. One conceivable example of this is international organizations where what gets to be discussed is always subject to contests of influence. By using one's influence to exclude certain issues from the agenda, one can affect the behaviour of actors in the international system. For example, by focusing on global action on climate change, one is diverting attention away from other issues, shaping the international system in the process. The third face focuses on one's ability to shape the other's preferences and interests in a way such that they align with one's own. This is the most constructivist dimension of power where the target is the core ideational concepts of the target. By reshaping how the opponent views itself and the world, naturally their behaviour will change, ideally in a desired way. The United States participated in the post-war reconstructing, and reshaping, of Japan. The introduction of Article 9 into Japan's post-war constitution came with American backing. This, and the post-war political discourse promoted in the country, strongly contributed to the contemporary anti-militarist sentiments of Japan, shaping Japan's behaviour in a way that was viewed as favourable by the United States, transforming a potential regional rival into an ally. These more indirect uses of power exemplify structural power: Where countries can utilize their position within the framework of the international system to exert influence over direct country-to-country uses of power. Essentially, Nye identifies these influences as soft power. The most classical example of soft power is culture: The attraction of one's own culture that encourages imitation in other countries, leading to the introduction of norms and values. The United States possesses a culture that is attractive abroad, with many seeking to imitate American lifestyles. But with the adoption of cultural signifiers, there is also a diffusion of norms and values, such as individuality or personal liberty. One interesting example of this is the "yankii" subculture in Japan that embraced some signifiers of American culture to an almost cartoonish level. But in the process, they also emphasize a rejection of Japanese social rigidity and acceptable social norms. Similarly, in the former Eastern Bloc the desire to imitate Western lifestyles was combined with the seeking of Western norms of personal liberty and democracy. The application of soft power focuses on one's ability to influence rather than force. It focuses on one's ability to achieve objectives without relying exclusively on coercive hard power.

As with the other theories discussed in this part, the normative promise of soft power is clear. Similarly to geo-economic statecraft, some have interpreted it as rending hard power instruments to be obsolete. This argument focuses on the idea that why should one undertake the costs and suffering of compellence through military power if objectives can be achieved through more subtle means. And just as with geo-economic statecraft, the key problem with this argument is that it assumes that soft power can always substitute for hard power, just because it is recognized to be more effective in some situations where hard power was previously used. Nye recognizes this and strongly cautions against it. He argues that he coined the term "smart power" exactly to recognize it as a fusion of soft and hard power, and to counter the perception that soft power alone is sufficient to conduct international affairs.[19] Nye argues that "smart power [is] the ability to combine hard and soft power resources into effective strategies [...] smart power suggests it is best to have both".[20]

One issue with the execution of a soft power strategy is that countries can be highly aware of such attempts to influence them. Western policy towards China has for a long time emphasized engagement to encourage China to integrate into the existing international order. However, while China has been willing to institutionally integrate into the system, Beijing has been highly concerned about the influence of foreign soft power. They view attempts to expose the local populus to foreign influences as an attempt to force "peaceful evolution" in China, where exposure to foreign norms will promote the downfall of the communist regime in favour of Western values of personal liberty and democracy. China has been pushing back hard against such attempts by controlling the flow of information and creating education campaigns such as the patriotic education campaign. Such efforts can limit the utility of subtler forms of influence, necessitating the application of other foreign policy tools. Once it became evident that peaceful evolution is not achieving the desired effects on Chinese behaviour, both the United States and Japan turned to more direct tools of power, emphasizing military deterrence in their foreign policy.

Another issue is the reliability and timeframe of soft-power instruments, which faces problems similar to constructivist approaches described in the next chapter. One can consider the issue of why a certain culture possesses global appeal while others don't and whether one can make a culture appealing. China has been seeking soft power for a long time. During the Hu Jintao era when its focus was on the "good neighbour" policy and countering emerging China threat theories, China was seeking to acquire soft power tools, including cultural influence similar to that of the United States, Japan, or South Korea. However, China so far had limited success in achieving such structural influence, despite its emphasis on Asian values and cultural performances such as the 2008 Beijing Olympics' opening ceremony. And even if the appropriate tools were acquired, translating those influences into direct policy outcome can be a significant challenge. Instituting the fundamental shift in an opponent's identity that the third face of power implies also takes an extended period of time. These tools best serve long-term political objectives and have limited utility in the case of immediate policy interests.

Overall, we can see that Nye is correct in his assertion that soft power, while important, cannot tackle all foreign policy situations. It needs to be used in conjunction with hard-power instruments, including the use of force, and as such it is reductionist to argue that soft power would offer a replacement for military power. Once again, we see a situation where the argument is for additional tools, the diversification of the toolset, but this does not inherently mean the discarding of existing tools. While geo-economic statecraft and soft power offer an opportunity to adopt a more nuanced foreign policy, military power and the use of force will continue to constitute a component of interstate interaction unless the policy situations where they offer utility are eliminated. Thus, declaring military power obsolete is premature.

Conclusion

While the theories outlined in the previous chapter have argued that the long-term negative consequences of aggressive behaviour normally outweigh the potential benefits of it, and thus it should be considered irrational to pursue, the theories in this chapter have highlighted alternate policy avenues to achieve the same geopolitical objectives without suffering the costs of war. Both geo-economic statecraft and smart power explore how to use other forms of power – economic or structural power, as well as influence – to pursue a state's interest. While these theories make a compelling case that often it is better to rely on these instruments rather than hard power, as the various scenarios in this chapter show, they cannot offer a total replacement for military power. It is important to note that neither Blackwill and Harris nor Nye would argue that these avenues of power were meant to do so. Unfortunately, in International Relations discourse, they have sometimes been interpreted as such, especially as fascination with newer avenues of power inevitably leads to attempts to discard older tools of statecraft.

The main reason as to why these tools are insufficient to replace hard power is rooted in the anarchic nature of the international system. One has to give credence to Mearsheimer's argument that force remains the *ultima ratio* of international politics. While economic and structural influence is important in the day-to-day interactions of international politics, they are all dependent and backed up by hard power to be effective. In an anarchic system, the opponent using force is always an option and thus it is necessary to maintain force to defend against hostile actions. On the flip side, if an opponent possesses sufficient power of its own to resist, it becomes all too possible that use of force remains the only option to effectively pursue a policy interest. As discussed in the previous part, hard power is largely unique in the state's policy arsenal for being able to bypass the will of the opponent in order to directly seize and hold objectives.

The next chapter moves beyond the assumptions of operating in a set international system and trying to best figure out its rules and the opportunities it presents and rather it ventures into social constructivism to explore the idea of fundamentally altering the nature of international relations to create a less competitive international system. If one could realize an international system where states, despite anarchy, do not view each other in competitive terms, then one could potentially eliminate the space for war in international politics since it is largely underpinned by the prevailing fear and struggle for survival in a hostile international system.

Notes

1 Nye J. S. (2011) *The Future of Power*. Public Affairs: New York.
2 Blackwill, R. D. and Harris, J. M. (2016) *War by Other Means: Geoeconomics and Statecraft*. Harvard University Press: Cambridge.
3 Ibid., pp. 24.
4 *The Constitution of Japan.* (1946). https://japan.kantei.go.jp/constitution_and_government_of_japan/constitution_e.html.
5 Wendt A. (1992) Anarchy Is What States Make of It: The Social Construction of Power Politics. *International Organization* 46(2), pp. 391–425.
6 Waltz, K. (2010) *Theory of International Politics.* Waveland Press: Long Grove.
7 Ibid.
8 Blackwill, R. D. and Harris, J. M. (2016) *War by Other Means: Geoeconomics and Statecraft*. Harvard University Press: Cambridge.
9 Schelling, T. (2008) *Arms and Influence*. Yale University Press: New Haven.
10 Morgenthau, H. (2005) *Politics among Nations: The Struggle for Power and Peace*. McGraw-Hill: London.
11 Nye J. S. (2011) *The Future of Power*. Public Affairs: New York.
12 Schelling, T. (2008) *Arms and Influence*. Yale University Press: New Haven.
13 Ibid.
14 Nye J. S. (2011) *The Future of Power*. Public Affairs: New York.
15 Ibid., pp. 19.
16 Corbett, J. (1911) *Some Principles of Maritime Strategy*. Guttenberg Online.
17 Gray, C. S. (2009) *Fighting Talk: Forty Maxims on War, Peace and Strategy*. Potomac Books: Dulles.
18 Nye J. S. (2011) *The Future of Power*. Public Affairs: New York.
19 Ibid.
20 Ibid., pp. 34–35.

7 Paradigm shifts

The previous two chapters focused on ideas about why states would rationally choose not to go to war. In contrast, this chapter examines two approaches to why states would not even consider going to war. The liberal tradition, represented by Mueller, focuses on humanity recognizing the horrors of war and leaving behind a less enlightened age where such methods would be considered acceptable avenues of statecraft. The second part of this chapter departs from the traditional dichotomy of realism versus liberalism to venture out towards the third large theoretical school of thought, that is, social constructivism. While often diametrically opposed to each other, both liberalism and realism share the assumption that the international system exists in some form independently from states and that the primary intellectual endeavour is to discover the rules of the system in order to formulate the best policy to survive and thrive in the system. For example, realism presupposes that anarchy is inherently threatening, creating competitive relationships between states by its very nature. In contrast, social constructivism argues that the meaning of key concepts in international politics is not predetermined but they are given meaning by the actions of states. Anarchy simply means an absence of higher authority; it is neither threatening nor welcoming by default. If the system is inhabited by states that act in an aggressive manner, then it becomes a matter of survival, just as if the system is inhabited by cooperative states then there is little to fear under anarchy.

Mueller and the obsolescence of war

In his book *Retreat from Doomsday,* John Mueller proposes a simple idea: "major war [...] may be obsolescent".[1] The idea that states have developed better tools for conflict resolution is a central component of this theory.[2] Thus, one could think that Mueller's theory would be better suited to the previous chapter on better tools available. However, Mueller's theory goes beyond this concept: At the core of it, Mueller argues that a fundamental shift has taken place in post-World War II international relations that has made major war unthinkable. Based on this, Mueller is better suited to be discussed in this chapter, which focuses on shifts within the framework of international relations.

The central thesis of this approach is not only that there are better tools available, but that contemporary major powers have collectively recognized war as wholly irrational. Mueller argues that "the notion has been gaining acceptance in the developed world that war is both abhorrent – repulsive, immoral and uncivilized – and methodologically ineffective – futile".[3] There are a few components to this that need to be unpacked to appreciate the thesis of the theory.

First, Mueller argues that the escalating costs of war have rendered it an unattractive option.[4] World Wars I and II have demonstrated that war among industrialized nations can occur on a previously unprecedented scale. The Franco-Prussian War of 1870–1871 has produced less than 200,000 fatalities. The American Civil War resulted in some 600,000 fatalities. In contrast, World War I produced some 20 million dead, while World War II resulted in some 70–80 million fatalities. Death has occurred on an industrial scale in these conflicts. Furthermore, they have displayed the scientific and industrialized brutality of modern war: World War I demonstrated the destructiveness of chemical weapons and mass infantry attacks, while World War II showed the tragedy of the use of nuclear weapons and the systemic extermination of entire groups of people. Under such conditions, especially with the proliferation of nuclear weapons, it seems almost unthinkable that there would be any objective that is worth the costs of modern war.

Second, in Mueller's view, the interests of states have fundamentally changed. Rather than focusing on hard power and national honour, modern states are primarily concerned about prosperity and wealth. These objectives are not only usually hard to pursue through hard power instruments, but war is often directly detrimental to them. Combined with the previous point, Mueller links up with Angell (discussed in Chapter 5) concerning the futility of war, both being part of the larger liberal tradition.

Third, in Muller's words, "war has lost the romantic appeal it once had, and it has been discredited as a method for obtaining desirable goals".[5] It is not only that states have determined war to be irrational, but more so that states reject war even as an option for consideration, war becoming "subrationally unthinkable".[6] From a moral perspective, developed states have "evolved" to the point where they would not consider war as an option to resolve their differences. It is not only that states would reject it during policy calculations but rather that simply it wouldn't occur to them to use it in the first place. Mueller represents the moral optimism of the liberal tradition that Morgenthau has dismissed as idealism: That the conflictual nature of the international system is a result of depravity and ignorance and with enough effort it can be corrected, allowing states to escape the horrors of war.

Overall, Mueller argues that the international system has reached a point where the idea of major interstate war is virtually unthinkable. Mueller in this regard invokes a familiar imagery. He compares the obsolescence of war to the abolition of slavery and the practice of duelling falling out of fashion.[7] This is the idea referenced earlier in this part that war is a relic of a less enlightened age that humanity will eventually grow out of as it continues its eternal progress. The idea that at one point humanity will come to discard it as a tool of statecraft, not because of specific political calculations, but because it is recognized as something unacceptable. Another apt metaphor to describe Mueller's approach would be comparing war with human sacrifices. All around the globe it has been a custom to sacrifice human lives to the gods for good harvest or to prevent a volcanic eruption. In Mueller's view the time has come to abandon the custom of sacrificing human lives at the altar of political ambitions, just as we ceased to do so to gain the favour of ancient agricultural gods.

It is needless to say that this book would disagree with some of these conclusions; otherwise, there would be little point to writing a book about making the rational decision to go to war. At the same time, statistical evidence also calls the growing peacefulness of the international system into question.[8] The basic argument of realism remains relevant: Competition, and the resulting bloodshed, is inherent to the structure of the international system. It is not motivated by failures in human thinking, but rather is a rational response to an anarchic international system. Mueller acknowledges this to a certain extent, arguing that competitiveness in the international system has not diminished, but that states have developed better tools to handle their political affairs, especially in a system where the primary focus is wealth creation.[9] This book has engaged with these ideas in the previous two chapters, coming to the conclusion that while there might be a diminished space for war, it is not gone. War continues to be the only reliable way to deal with an uncooperative opponent whose will cannot be broken. While humanity might have found a particular arrangement of the international system that has diminished the occurrence of war, this is inherently temporary and war would return as soon as the conditions of the international system have changed. And the international system evolves continuously. And while wealth is an important consideration, it is not the only one. In envisioning the potential return of war, Mueller argues that the most likely trigger would be a shift in the international system away from emphasizing prosperity.[10] The capitalist world economy continues to be highly focused on wealth; however, we can see cracks in this thinking as it sinks into ever deeper systemic crisis.[11] The sustainability of wealth creation is a central question of contemporary politics. The materialist focus of the late Cold War period and the years following the collapse of the Soviet Union is increasingly replaced by one focused on other values and identities. It is no longer about being rich, but about creating a liveable future. Correspondingly, in development studies human security is increasingly replacing economic development as the main focus.

The logic of rejection of war worked in the bipolar logic of the Cold War. Mueller acknowledges that one the reason for these developments has been that both of the superpowers have been largely satisfied with their sphere of influence.[12] Conditions since then have changed significantly. The collapse of the Soviet Union has led to the creation of a unipolar order. A rising China is now pushing against this order, seeking to carve out its own sphere of influence, which is opposed by the United States at every turn. China has long been viewed as a pragmatic country primarily concerned about wealth creation, which would seem to make it fit into Mueller's model. But this ignores China's larger political ambitions as well as the Chinese Communist Party's primary focus on regime survival. China is entering the stage where is seeks to translate economic influence into actual political power. It views becoming a great power as fundamental to its security. At the same time, the United States has shown an unwillingness to give up its unipolar grip on the system. President Obama created his pivot to Asia to directly contain a rising China and to preserve the status quo. This renders the system a pressure cooker: Tensions between the US and China will keep rising as they push against each other. Unless one of them gives way to the other, a violent explosion is a potential outcome. Unlike the bipolar Cold War order, the current system is inhabited by fundamentally unhappy major powers: China seeks to ascend to great power status, Russia wishes to recapture its Cold War relevance, the US grapples with its potentially waning influence, Japan seeks to carve out a more active role in regional security, Great Britain is seeking to break away from Europe, and the European Union is re-evaluating its ties with the United States and its role within the existing order. It is common to highlight China as a revisionist power. But fundamentally, many of the major powers are unsatisfied with the status quo. And their political ambitions put them on collision courses.

A fundamental failure of this approach is assuming that peace will become self-enforcing by war becoming subrationally unthinkable. Humanity has achieved a period of peace because of the horrors of war and nuclear annihilation posed a stark existential threat. War did not become obsolete, but rather states adopted rational calculations and pursued effective preventative diplomacy. War can be prevented. As this book will show, by altering rational calculations one can manoeuvre to a position where war can be deterred or channelled into more constructive avenues. But then peace is not a result of the obsolescence of war. Rather it is the result of the fear that war is anything but obsolete.

Wendt and the construction of international politics

Alexander Wendt argued that anarchy does not have an inherent meaning, unlike other major schools of thought that considered such a meaning to be given. For example, realism considers competition and self-help to be inherent to anarchy. Wendt's social constructivism argues that the international system is part of our social reality rather than our material world. The international system is not naturally occurring and no element of it has inherent meaning. Rather the actors within it create meaning, primarily through their interactions and shared history.[13] This is an intriguing possibility as such an approach does away with the determinism of other schools of thought. Most importantly for this chapter, if the meaning of anarchy is constructed, then the fundamental aspects of the international system can be redefined, allowing us to escape realist power competition. Or at the very least, that is one of the normative implications of theory. Unfortunately, as seen already above, such normative implications are rarely realized. This section will focus on two main issues: First, it will examine if social reality is constructed through interaction, and then how does waging war affect this. This has already been partially covered in Chapter 3. Second, the section will explore the possibility of altering competitive dynamics in the system through constructivist policies.

At the centre of Wendt's social constructivism is identity and interaction. Every state has an identity based on its culture and history. This identity affects state interests and behaviour in the international system. One good example to illustrate this is the difference between how China and the United States views security. China has a long contemporary history of territorial violation: imperial powers form Europe gradually dismantled China's sovereignty during the 19th Century, extending their control from the coastal provinces through superior military might. Similarly, China's 20th century history has been shaped by several wars with Japan, including a full-blown invasion, and direct territorial threats from the US and the Soviet Union. Unsurprisingly, China defines its security interests in strongly territorial terms. This can be observed in China's focus on gaining physical control over the East and South China Seas by establishing forward outposts. In contrast, the US security emphasizes such direct homeland defence much less, being a country surrounded by only two weaker states and separated from all other major powers by oceans. The US does not have a history of invasion comparable to China; thus, it defines its security interests and postures differently. This divergence in identity is further underpinned by the US being a global superpower, which inherently affects its security outlook. US security policy puts a bigger emphasis on ideology and rule-setting over strictly territorial considerations. This is not to deny a geopolitical component to US security. But theories such as Barnett's core-gap thesis, which was a good representation of contemporary US strategic thinking, focused on issues such as connectedness, globalization, and setting global rules, instead of natural boundaries and maritime exclusion zones. Put simply, China's security posture is much more defined by a survival imperative, compared to the U.S. that emphasizes global control and leadership. It would be reductionist to ignore such a difference in identity and outlook and argue that both countries' security interests and policies can be understood using the same extrinsic security interests as foundation.

Henry Kissinger used weiqi (or go) and chess to illustrate a similar point.[14] Kissinger argued that Western strategic behaviour is based on the logic of chess: Players seek out engagements with the express purpose to remove the opponent's pieces from the board. War fits into this context as straightforward contest of power with clear winners and losers. In contrast, he argued that Chinese strategic culture is closer to weiqi, where the focus is on gaining situational advantages over time, rather than removing pieces from the board. Correspondingly, Chinese strategic behaviour emphasizes political objectives, positioning and psychological factors. The PRC does not necessarily goes to war to win on the battlefield, but to seek a psychological shift that results in political changes.

Identities are neither inherent to the state, nor are they static. A central component of Wendt's constructivist argument is that identities evolve through interaction. One good way to illustrate this point it to consider the fundamental philosophical question of whether humans possess innate programming. Abrahamic religions often view humans as inherently sinful. A similar thinking has been expressed in traditional IR schools of thought. Realism and liberalism assume that states have inherent interests essentially hardcoded into the state that animate it. Realism's focus on security is such a hardcoded interest. This inherently affects their behaviour, rendering them suspicious and fearful. In contrast, Wendt's theory is closer to Locke's concept of *tabula rasa*, meaning that that states do not possess such inherent programming but their behaviour evolves as their identities change through interactions. States are viewed as essentially learning entities that base their behaviour on a history of experiences instead of hard-coded factors.

One good example of this is Japan. Japan's identity has shifted significantly and this has affected its behaviour more so than extrinsic security seeking could explain. During the Tokugawa Shogunate Japan adopted an isolationist stance, disengaging from the world. This was shattered by the arrival of Commodore Perry's "black ships" after which Japan quickly shifted its posture; the Meiji reform period adopted Westernization and industrialization, fundamentally altering Japan's identity. Japan sought to imitate the Western imperial powers, who demonstrated a successful model of development, leading Japan to develop its own imperial ambitions. These ambitions culminated in the Pacific War and the nuclear bombing of Hiroshima and Nagasaki. Once again, Japan shifted its identity, adopting anti-militarism and focusing its foreign policy on economic engagement. This leads up to the growing animosity with China that once again pushes Japan to redefine its own identity, forgoing anti-militarism in favour of regional leadership and direct deterrence. Naturally, in each of these instances security seeking played an important role: Japan was responding to external threats each time it majorly redefined its identity. But at the same time, one cannot argue that Japan's interests and identity remained static. Each response to such external stimuli fundamentally altered how Japan viewed itself, its role in the international system, and the very definition of its own security and how best to achieve it. This evolution is beyond the scope of traditional International Relations theories such as realism. This lends credence to Wendt's argument that anarchy and states' response to it is constructed instead of pre-defined.

International relations is a collection of interstate relationships with their own ideational considerations. The international system is the intricate web of state interactions and the identities that underpin those interactions. And as these evolve, the international system is in constant flux as states define and re-define their interests and relationships. Wendt identifies three basic patterns of relationship that underpin the system:[15]

- Cooperative: There is a positive association between the security of two states.
- Neutral: There is no association between the security of two states.
- Competitive: There is a negative association between the security of two states.

A competitive relationship is most reflective of realist thinking: Two states view each other as threats and engage in balance of power politics to prevent one or the other gaining an upper hand in the relationship. A good contemporary example is the relationship between China and Japan. As China pushes for more power, Japan seeks to push back to maintain an equilibrium. Similarly, when Japan seeks to advance its regional leadership, China is the vanguard of the opposing force. In contrast, in a cooperative relationship power gains made by one are seen as positive for both, more reminiscent of the positive-sum thinking of liberal International Relations theories. An example of this would be the relationship between Japan and the United States. Japan does not view power gains by the United States as inherently threatening. American involvement in the Asia-Pacific is viewed as complementary to Japanese interests, thus Japan sees them as mutually beneficial. Similarly, the United States provides active assistance to increase the power of Japan in the region based on the same recognition of shared interests. A neutral relationship simply means that states don't attach importance to their security relationship with each other. An exaggerated example would be Moldova and Nepal. Neither is concerned about the power moves of the other and they neither gain security nor insecurity from these moves.

The main normative promise of the theory is based on these two principles: States can have a range of associations and their relationships can evolve over time. It is reasonable to assume then that states can purposefully evolve their relationships in positive ways, seeking to shift them from competitive to cooperative stances, essentially ending the competitive power dynamics that plague international relations and so often lead to conflicts. Again, Japan presents a good example. Prior to World War II, Japan's relationship with the U.S. devolved into competitive dynamics. Japan viewed the U.S. as fundamentally a threat to Japanese interests. After the war, the U.S. has occupied Japan. However, this did not lead to further animosity, but rather through the post-war cooperation of the two states a cooperative identity emerged. One could seek to dismiss this as an alliance of convenience during the Cold War based on the realist maxim that there are no permanent friends or enemies, just permanent interests. However, this does not capture the value Japan attaches to its cooperation with the U.S. or how the relationship continues to this day, despite expectations that as Japan's economic power grows in the 1970s and 1980s it would devolve into a power contest. Similarly, the previously war-torn continent of Europe was unified under the European Union, ending centuries of competition between France, Germany and the United Kingdom. While there are conflicts of interests, there is a high value attached to the relationship that prevents the devolution of relations into purely competitive power politics. France is not suspicious of Germany's military as a tool of invasion. On the contrary, the EU is seeking the creation of a common military force. Thus, there is merit to the normative argument that a foreign policy focusing on the purposeful positive evolution of relationships could lead to a more harmonious and positive-sum international system.

This is where one can also argue that war doesn't work, especially in the long run. In a system of relationships, especially dominated by imperfect information, reputation is key. When deciding how to deal with other states, any state would look at not only their own shared history, but how a state treated others. When deciding whether to put trust into a neighbouring country by evaluating their security relationship neutrally or cooperatively, a state would unavoidably look at whether that country betrayed others' trust before. If the state has a reputation to break deals or use its military power offensively, it would find it extremely difficult to engage in a constructivist strategy based on trust. In contrast, if the state has a reputation for honouring deals or benign conduct, it would be easier for others to trust that state. Japan has encountered this problem with South Korea. Both countries are part of the American-led alliance system in the Asia-Pacific and their fundamental strategic interests often align. Yet, South Korea has been hostile to Japanese attempts to elevate its own regional security profile, demonstrating that their relationship is neutral at best, and often competitive. South Korea does not trust Japan to be a benign regional power due to their shared history of invasion and colonial subjugation. Importantly, South Korea is not satisfied that Japan's current conduct demonstrates sufficiently that it has altered its identity enough not to pose a threat to the country. Engaging in the offensive use of force will create a reputation for a state as a "warmonger" and will increase threat perceptions. Thus, even if war delivers benefits in the short term, in the long term it makes cooperation more difficult as states will increasingly adopt a competitive stance towards the aggressor, even if they are not the target of the aggression, in this sense constructivism offers a conclusion similar to Waltz's defensive structural realism. A system organized around history, and where military aggression is viewed as a violation of common norms, waging wars presents significant costs. On the flip side, among states that view the use of force as a virtue or have similar ambitions, engaging in aggressive behaviour can foster better relations. One only needs to think of the alliance between Germany, Italy, and Japan in the 1930s and 1940s. However, whether there is any stability to such relations in the long term remains dubious, as one could reasonably expect that they would eventually turn on each other for dominance.

While Wendt's constructivism offers great normative promise, in execution it unfortunately remains more limited. As with many of the theories discussed above, pragmatic execution encounters several obstacles. Changing the dynamics between two countries is neither easy nor a short process, as illustrated by the aforementioned example of Japan and South Korea. There are a few considerations that limit the feasibility of this strategy.

First, there are no do-overs for first impressions. The system has been in existence for millennia now. Each state has a long history of interactions and it can be difficult to overcome such entrenched identities. For example, China has over 5,000 years of civilizational history that informs its behaviour, underpins its interests, and influences who China views as friends and who as enemies. Overcoming that to radically alter the dynamic between two countries – to shift from competitive to cooperative – would be a colossal undertaking. This is especially so as long histories for both competitive and cooperative behaviours are self-reinforcing. Breaking the vicious cycle is extremely difficult. Even Wendt recognizes this when he argues that, although anarchy does not have a set meaning as realism would argue, it is likely that the current system can only operate under the realist conditions due to the entrenchment of power politics as a behaviour.[16] A competitive dynamic is especially difficult to break as such a shift would require trust, which is in short supply between competing states.

Second, it is hard to determine how to adopt a policy that reliably shapes shared identities and does not simply create alliances of convenience that break down once external constraints are removed. The United States has been successful in building a cooperative relationship with Japan. It had a similar opportunity with China: After the Sino-Soviet split in the 1950s and 1960s, China has shifted its allegiance towards the American-led alliance system. In 1972 Nixon visited China, beginning the process of rapprochement with the People's Republic of China. However, the re-establishment of cooperation was not followed by the same shift in the relationship as with Japan. Contrary to that, once the Cold War ended, the United States and China returned to a largely competitive dynamic, rendering their relationship during the Cold War merely an alliance of convenience. One can point to many factors for this, including the US continued engagement with Taiwan, the identification of China as the new target of post-Cold War strategic planning, China's own strategic designs for the region, and so on and so forth. But the difference illustrates the difficulty of purposefully designing a strategy that works in such a constructivist fashion. Relationships are intangible and seemingly mundane events can have unforeseen consequences as they are filtered through the perceptions and cognitive biases of the various states. The evolution of interstate relationships seems to have a mind of their own and affecting purposeful change can be challenging to nigh impossible.

Third, the difficulty of devising effecting policy is further compounded by the fact that shifting a competitive relationship to a cooperative one often requires the wilful acceptance of insecurity, violating basic strategic logic. The viciousness of a competitive cycle is often underpinned by the reciprocal moves to match power increases (as recognized by the realist security dilemma). Breaking this cycle would require one state to abandon such a mentality in order to signal to the other that it desires a change in their shared relations. However, such purposeful inaction means the acceptance of an unfavourable shift in the balance of power without any guarantee that the opponent will reciprocate or is willing to engage in de-escalation. Essentially, such a policy is inherently risky for any state. There are few examples when states accepted limits on their pursuit of power, but mostly when the consequences would have been devastating otherwise (e.g. limitations on chemical, biological, or nuclear weapons). In the latter days of the Cold War, the United States could afford to give some concessions to the Soviet Union, because they both possessed enough nuclear arms to end the planet several times over. This is not an option to many other powers.

Finally, the target regime has to be open to de-escalation. This might not be the case always. While war is universally understood to be terrible, there are political benefits to animosity. Both China and North Korea have nationalist political discourses that fuel a competitive dynamic with the United States. In both countries these narratives are central to regime survival and thus no amount of engagement will significantly alter them. Under such conditions there is hardly any other choice than hard-power-based competition, where the use of force might be necessitated to protect key interests. It is good to try to get along with others; however, this is not always possible. Sometimes an opponent is not interested in the changing of hearts and minds. China especially is a problematic case. Beijing's threat perceptions are magnified by Western efforts of engagement which it sees as attempts to bring "peaceful evolution" leading to the downfall of the communist regime. Thus, in the case of the PRC, while Beijing is happy to reap the benefits of cooperation, it is highly resistant to any attempt to alter its identity. Even in the case of Japan, a military defeat was necessary before substantial shift in the relationship could take place.

Conclusion

Overall, we are encountering the problem that keeps repeating throughout Part II: While various theories can successfully argue why the occurrence of war is reduced in the contemporary system, they struggle to provide an iron-clad argument for why war would become obsolete. The fact remains that war continues to occupy a unique position in the state's foreign policy arsenal and the structure of the international system remains such that force as a tool remains useful. Mueller's argument for the obsolescence of war is held back by both the limitations of other alternative tools and that the moral rejection of war runs contrary to the anarchic nature of the international system. Survival did not diminish as a key interest in such a system and the theoretical possibility remains that it might necessitate the use of force, even if for a period peace has been achieved. Two ideas that come to mind are both tied to revolutions in military affairs: Opposing states could develop defence technologies that would render nuclear weapons ineffective or the increasing automation of war through the adoption of drones could reduce the deadliness of combat. Both would strike at the core of states' supposed enlightenment about the abhorrent and unacceptably devastating nature of war. Arguably states rejected the method of war, industrialized total war that has characterized the 20th century, rather than the idea of war. With the evolution of warfare it remains possible that combatants will once again resort to the use of force if the costs are acceptable.

Similarly, social constructivism fails to show how the nature of the international system could pragmatically become fundamentally different. After thousands of years of human history, the competitiveness of the system seems thoroughly entrenched. While Wendt's constructivism can offer another tool for the state's foreign policy arsenal, it cannot unequivocally replace the use of force. The theory has significant normative promise. But arguing that it renders wars obsolete is a bridge too far. One case study to illustrate this is to return to Japan's post-war foreign policy. As discussed above, Japan has primarily engaged in economic diplomacy (see above) while forgoing direct military might. But Japan's policy goes beyond simply engaging in economic interdependence or geo-economic statecraft. There is a strong ideational component, both how Japan views its own role in the contemporary international system and how it seeks to build relationships with others. Japan's economic diplomacy is a policy of integration, not simply interdependence. As discussed above, Japan often employs economic tools in its foreign policy. But there is a larger ideational component to it. Japan's ultimate goal has been regional economic integration through its "flying geese" paradigm. While the integration of supply chains and diffusion of technology were the tangible backbones of this policy, they also carried with them an implication of the diffusion of Japanese norms and soft power influence. Japan viewed such cooperation as the ultimate backbone of regional stability and was willing to engage in cooperation and official development assistance even when the recipient country adopted a hostile political tone. Japan was highly conscious of the drawbacks of isolating China in 1989 and even as their political relationship grew antagonistic, Japan continued to provide assistance to China in order to facilitate its constructivist aims. China remained one of the largest recipients of official development assistance (ODA) well into the 2000s, despite growing antagonism between the two countries. Japan's objective has always been to "socialize" China into the international order to ensure peaceful coexistence. However, despite all of its efforts, Japan's policy has ultimately failed. China remains antagonistic to Japan and tensions between the two countries remain high. China continues to view Japan in highly competitive terms. Tokyo was eventually forced to alter its policy course and to return to seek traditional hard power deterrence against Beijing, highlighting the limits of a constructivist approach to security.

Notes

1 Mueller, J. (1990) *Retreat from Doomsday: The Obsolescence of Major War*. Basic Books: New York, p. 240.
2 Fettweis, C. J. (2006) A Revolution in International Relations Theory: Or, What if Mueller Was Right? *International Studies Review* 8(4), pp. 677–697.
3 Mueller, J. (1990) *Retreat from Doomsday: The Obsolescence of Major War*. Basic Books: New York, p. 217.
4 Ibid.
5 Ibid., pp. 220.
6 Ibid., pp. 240.
7 Fettweis, C. J. (2006) A Revolution in International Relations Theory: Or, What if Mueller Was Right? *International Studies Review* 8(4), pp. 677–697.
8 Braumoeller, B. F. (2019) *Only the Dead: The Persistence of War in the Modern Age*. Oxford University Press: Oxford.
9 Mueller, J. (1990) *Retreat from Doomsday: The Obsolescence of Major War*. Basic Books: New York.
10 Ibid.
11 Wallerstein, I. (2006) *World System Analysis: An Introduction*. Duke University Press: Durham.
12 Ibid.
13 Wendt A. (1992) Anarchy Is What States Make of It: The Social Construction of Power Politics. *International Organization* 46(2), pp. 391–425.
14 Kissinger, H. (2011). *On China*. Penguin: London.
15 Wendt A. (1992) Anarchy Is What States Make of It: The Social Construction of Power Politics. *International Organization* 46(2), pp. 391–425.
16 Ibid.

Part II

Conclusion

The obsolescence of war

Part II started with a simple premise: There are several theories in International Relations scholarship that argue that war (use of force) and military power face a diminishing importance in International Relations. This is true to a certain degree. States possess a diverse foreign policy toolset; military power is simply one of the tools in it. Geo-economic statecraft and soft power often offer better utility and it would be intellectually dishonest to argue otherwise. Power is neither a simple nor a one-dimensional concept and its exercise needs to be agile. But this does not mean that one can fully discount military power. As the various examples in this part have shown, just as military power has its limits, so do all other foreign policy tools. Good policy is based on selecting the appropriate tool. And sometimes it is necessary to wield the sword of war. If an opponent is resistant to other forces of influence and the interests at stake are vital, then one has no choice but to fight. Discounting this possibility is analytically unhelpful. There is an obvious, visceral opposition to war. Such opposition should exist. However, such opposition should not cloud one's rational policy judgement.

The main point to take away from this chapter is that force is a situational tool. One is required to identify the situation in which it can be correctly used. While this part has focused on arguments against armed confrontation, the next part explores the reasons for which one goes to war. As such, the next part examines the various theoretical approaches that explain the occurrence of war. Building on the combination of this and the next part, one can build the foundations for the rational choice to go to war. Even if one does not wish to go to war, one must understand the process of war in a rational fashion to anticipate the moves of the opponents. Only when considering war seriously in a policy context can we have any chance of devising policies that can prevent needless bloodshed.

Issues for further thinking

This section offers a few questions and scenarios that one can explore to further reflect on the key points made in this chapter. For each question, this section provides some key details and points to consider but ultimately leaves the question open-ended.

There are states that do not possess military power. Does this not demonstrate that it is possible to succeed in the system without military force?

There are a number of states that possess no formal military of their own, including Andorra, Dominica, Grenada, Kiribati, Palau, Costa Rica, Iceland, Mauritius, and Panama. These countries nevertheless remain sovereign states that participate in the international system. One could reasonably ask whether their continued existence is not proof that realism is overly pessimistic about the system and that it is fully possible to survive, and on occasion thrive, in the system without military might.

There are a few caveats to these states:

- All of these countries are minor powers and small states. They would possess limited military power in the first place, if understood based on the principles of Mearsheimer due to their low latent power (small population and often limited economic resources). Kiribati has a population of some 115,000 people, with a gross domestic product (GDP) of some US$250 million. Iceland has a population of some 360,000. Assuming that only a small fraction of the total population could be devoted to active military service, they could only field a small force that would make little difference in a potential confrontation with a major power. Due to this they might also not devote their often limited resources to it.
- Most of these countries rely on external security guarantees from major powers that assume responsibility for their national defence. Andorra is protected by France and Spain. Dominica and Grenada rely on the Regional Security System, which is an international agreement to guarantee the security of the Caribbean region and involves assistance from the United States. Iceland has mutual defence ties with the United States and the latter maintains security guarantees for the country. As part of the North Atlantic Treaty Organization (NATO), Iceland also benefits from various collective security arrangements. Kiribati is protected by Australia and New Zealand. In essence these countries can forgo investing into national defence because major powers engage in status quo behaviour and deter any attempt to pursue a military confrontation between many of the smaller states within their orbit.
- While many of these states do not possess a formal standing army, they maintain semi-militarized or paramilitary forces as part of their police or domestic security forces. Costa Rica maintains its own *public force* with limited military capabilities. Iceland maintains a militarized Coast Guard, air defence systems, and other military-capable units. Mauritius maintains the paramilitary *special mobile force* and a coast guard.

The question one has to ponder is whether states reliant on external security guarantees are proof of the success of non-military strategies. While sovereign states, their foreign policy options are limited by their dependence on external security guarantors. And in an anarchic international system, they largely put their faith into (a) offering no reason for anyone to bother to invade them and (b) that if a crisis occurs their security guarantors will keep their word despite the anarchic nature of the system. Realistically, many of these countries would have limited chance of survival in the international system if not for limited strategic value and the goodwill of major powers that are interested in preserving the status quo. Nevertheless, they offer an interesting scenario in which a state without formal military power can nevertheless exist in the international system. One is encouraged to ponder whether it is possible to translate their experiences to larger nations and whether a largely disarmed international system where security is guaranteed by a few major powers could be accomplished. Would such a system offer peaceful coexistence? Or would such a system gradually descend into the tyranny of the security guarantors?

Should the United States resort to the use of force to accomplish denuclearization in North Korea?

If one understands war as a highly situational tool, then the question is when it is appropriate to use. When does war become unavoidable? One can see through the theories above that there are several alternate avenues to accomplish policy objectives. But the chapters above have also demonstrated that these avenues have limits, leaving space for war in foreign policy. The difficult part is to determine when alternate avenues have truly failed. War is a high-cost option and as such policymakers should be certain that they have exhausted all other policy avenues before deciding to pursue it as an option. But it can be difficult to know whether an alternate option, for example, geoeconomic statecraft, has failed because of poor execution, necessitating a change in execution, or because it was inappropriate to achieve an objective altogether.

The United States has declared non-proliferation as a key policy objective and has pursued it aggressively, whether in the case of Iraq, Iran, or North Korea. In the case of North Korea, the United States has been eager to achieve an end to Pyongyang's nuclear programme. To achieve this, the United States has put a lot of pressure on the regime in the form of economic sanctions, disrupting the regime's traditional alliance with China, and engaging in threats of violence in case of non-compliance. Yet, while Pyongyang on occasion appeared to be open to a resolution, the United States has so far failed to secure a decisive end to its nuclear programme. North Korea continues to stubbornly defy international sanctions through illegal means of sanction busting and through the callous disregard the regime displays for ordinary North Koreans. Arguably, the punitive measures failed to persuade the regime to abandon their nuclear ambitions and are only in place to spread misery among the already oppressed population that has no direct influence over the country's military projects. It seems unlikely that economic and political pressure on its own will be sufficient to break Pyongyang's nuclear ambitions, especially as the acquisition of nuclear weapons fits well into the regime's bounded-rational cognitive framework, the need for such weapons being confirmed by the US aggressive posturing.

This raises the question that, if the United States considers non-proliferation as a major national interest, then it should accept the failure of its diplomatic approaches and escalate its policy approach to open warfare. This is an especially pressing question as with each passing day North Korea is closer to acquiring a working nuclear arsenal, which will render military action virtually impossible. On the flip side, if the United States is unwilling to take military action over the country's nuclear programme, then can it credibly claim that non-proliferation is a major national interest, and through that sufficiently justify the suffering its policies inflict on ordinary North Koreans?

The primary question to consider here is whether one can reliably determine if the US diplomatic approach has failed. One has to critically examine whether more sanctions and more inflicted hardship would push the regime towards the desired policy stance. It is debatable whether Pyongyang possesses a breaking point that the United States could target reliably with its own foreign policy, especially as it only enjoys unenthusiastic support from China, which would allow it to claim a diplomatic victory. This will determine whether the failure of US policy is a result of poor execution or the use of inappropriate tools. One has to face the uncomfortable idea that if the United States is serious about non-proliferation, then the use of force might be the only scenario to successfully accomplish it. At the same time, if the United States is unwilling to go all the way to prevent a nuclear North Korea, then the suffering imposed ultimately serves little purpose beyond slowing down the regime in the name of theatrics.

Should one engage in projects with countries that pose a potential strategic risk?

Whether one can work together with competitors is a fundamental question for International Relations, especially if the anarchic international system is understood in a realist sense. Interdependence is a complex question: On the one hand, it can confer significant benefits and opportunities to create a peaceful international system, but on the other hand it might reduce the security of the country by creating unfavourable power relations and dependencies. Self-reliance is the ultimate form of security according to realism. However, it is debatable whether that is realistically possible. This debate has already been confronted by the 18th and 19th-Century economic thinkers. Mercantilism viewed economics in purely competitive terms, emphasizing self-sufficiency and aggressive trade policies demanding a positive trade balance. Anything less would endanger the sovereignty of the state. It was confronted by Smithian liberal economics that argued that its paranoid competitiveness was counterproductive and left the entire system poorer by disallowing the effective utilization of natural advantages and division of labour. With the downfall of mercantilism, free market economics began dominating international trade, leading to the creation of exploitative dependencies and an oppressive division of labour for many, reflected by the rise of Marxism and development studies in the following decades and century.

Cooperation offers an opportunity to fundamentally shift a competitive relationship into a cooperative one through mutual benefit. However, there is also a possibility of strengthening an opponent, negatively affecting the balance of power. Mearsheimer (as discussed in Part III) argues that one should not engage in win-win cooperation if somebody else benefits even more. The difficult question then is whether countries, in light of complex interdependence, should engage in cooperative behaviour with other states that could be a major strategic competitor. The Cold War has been a binary period: It is characterized by zero-sum competition between the United States and the USSR. Whenever one benefitted, the other suffered. At the same time, arguably one of the key things that prevented terminal escalation into total thermonuclear war were the instances where the two superpowers reached out and cooperated, de-escalating their relationship. For example, both accepted limitations on certain types of weapons, rather than seeking to turn them into a competitive advantage. Another contemporary example is the Belt and Road Initiative (BRI). There is a lot of anxiety on whether it is a geo-economic tool designed by China to entrap developing nations in debt and thus allow the spread of Chinese influence. This led some to argue that developing countries should forgo participation in BRI. At the same time, BRI offers critical infrastructure to these countries with strong financial support, a deal that is hard to turn down.

The international system is competitive overall, but that does not mean that it is competitive in any and all instances. Refusing to cooperate is just as much of an oversimplification as always trusting others to have our best interests in their minds. Finding the right balance can be very difficult and the costs of miscalculation are dire. But that is the cross policymakers bear. It is important to be critical to any offer of cooperation, but it is equally important that one does not become so cynical or paranoid that they let a good deal pass by.

Part III

Reasons for going to war

Part II examined theories that offered a normative argument concerning the obsolescence of war. In each case it was shown that, while the use of force is a situational instrument, no option offers the total elimination of military power in the international system. The use of force has been part of the international system for a long time and it continues to do so. The question then becomes of when force is being used in the international system. This part shifts the focus to examining various theoretical approaches that seek to explain the occurrence of use of force in the international system.

There is not a singular answer to why a state decides to go to war. Some approaches boil it down to singular factors, such as realism that focuses on the balance of power, but these are often simplifications of a more complex question. Yes, states pursue interests and the balance of power determines whether they can do so through military means. However, what those interests are, including what a state's security interests are and whether those interests are viewed worthy of military intervention are affected by a multitude of factors. One can argue that there are four levels of interests: Survival, vital, major, and minor. In this ranking military force is needed for the first two, maybe needed for the third and not suitable for the fourth. While this is normatively useful, to be analytically effective one needs to understand a lot of different factors. Whether a disputed territory falls under survival or major or minor interest is determined by a wide range of factors, and even then, it is often affected by seemingly subjective considerations specific to the bounded rationality of a certain state. Understanding what is at stake is essential when one makes the decision to fight.

This is even more so if one is to develop a preventative foreign policy. It is possible that one's own state is not interested in war. That does not mean that the neighbours do not harbour hostile intentions or feel forced to resort to the use of force. Understanding the key motivating factors of military confrontation can help recognizing these trends in the behaviours and politics of other states, in which case an effective foreign policy can be devised to deter them from a more destructive path. Thus, even if one is not interested in war, successful peace requires an intimate knowledge of the anatomy of war.

Whether for peace or for war, and whether to know ourselves or others, this part examines key approaches to the occurrence of war. The first chapter opens with a discussion on systemic factors. The following chapters discuss additional considerations, such as geography, culture, identity, globalization, escalation of territorial disputes, or domestic political considerations that can influence the occurrence of war in the international system. While realism provides a stable analytical foundation, a more comprehensive approach is needed, especially as structural realist approaches tend to be purposefully reductionist, even if readers sometimes neglect the nuance of this reductionism. The previous part discussed why one should not go to war. The main conclusion of Part II is that it is impossible to fully eliminate war from the system. Part III, accordingly, focuses on why one actually goes to war. War is terrible, and as such, it should not be undertaken without a good reason – maybe not a good reason from a moral perspective, but at least an intellectually grounded, rational one.

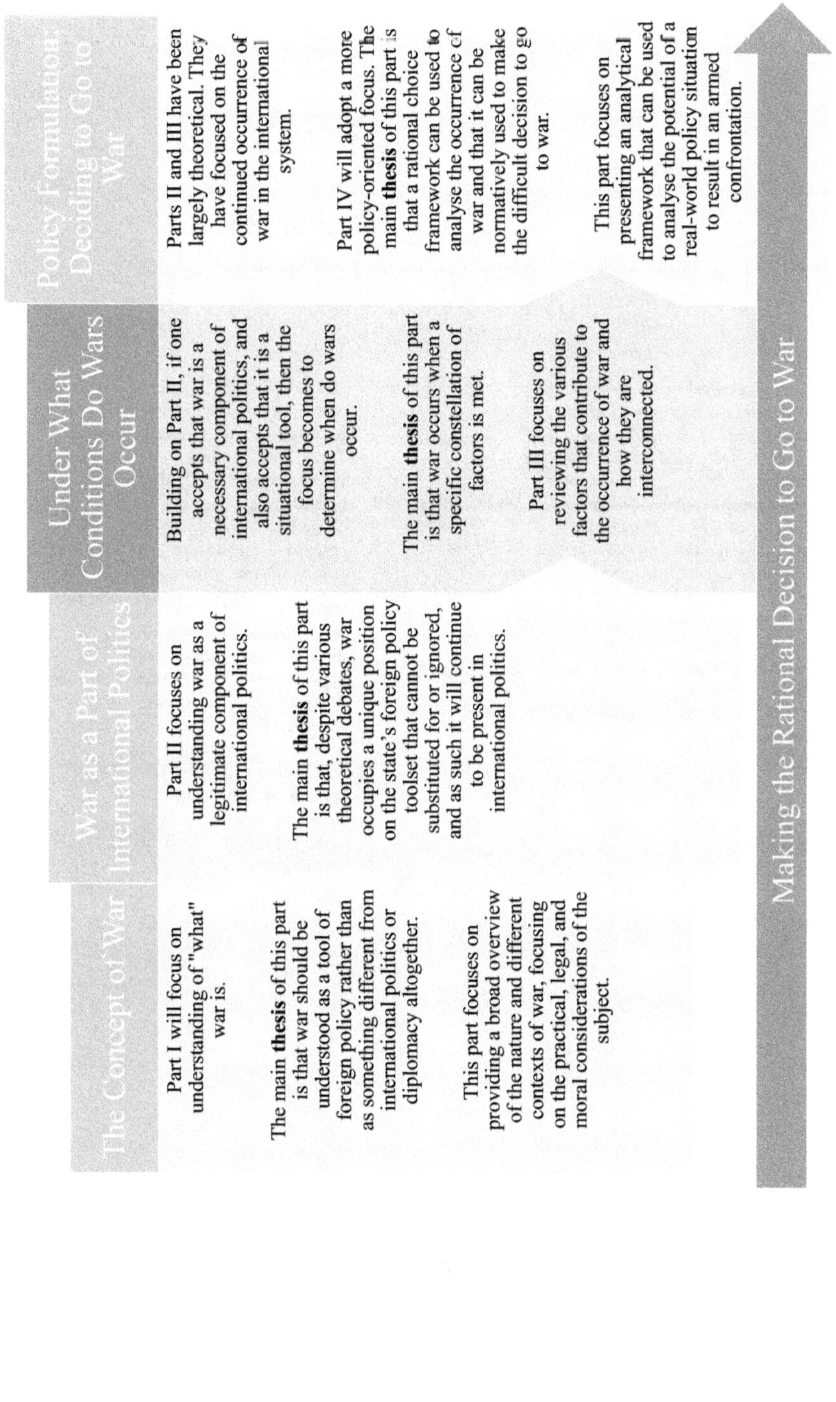

Figure PIII.1 Part III's place in the overall arch

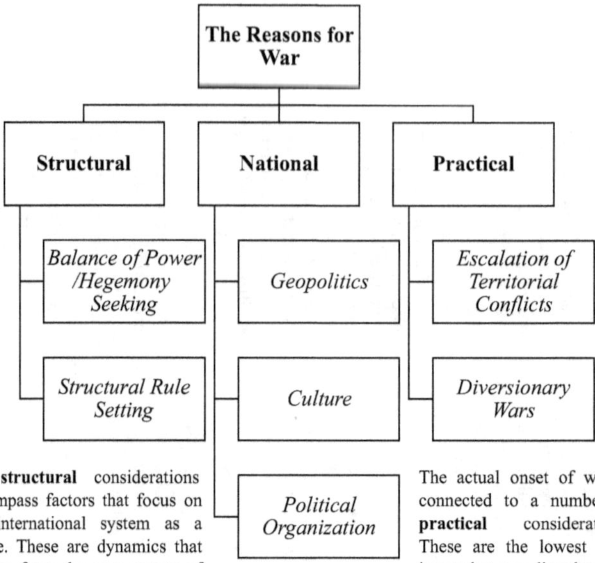

The **structural** considerations encompass factors that focus on the international system as a whole. These are dynamics that emerge from the very nature of international politics, such as balance of power considerations or setting the rules of the entire system in order to gain structural power. These are the highest-level factors that underpin state behaviour but on their own are often insufficient to explain the onset of hostilities. Rather they act as conditions that permit or incentivize war.

The **national** reasons for war focus on factors that are specific to a particular nation and thus are at a lower level than structural factors that encompass the entire international system.

The actual onset of war is connected to a number of **practical** considerations. These are the lowest level issues that may directly affect behaviour on the ground. These tend to be day-to-day political calculations or situations on the ground.

On their own they are insufficient to cause wars. In most instances they need higher level factors to create a fertile situation for war. However, they are important in explaining why a particular war, which may have been brewing in the background for a long time, has occurred at any given time and not another point in time.

Figure PIII.2 Structure of Part III

Table PIII.1 Key works for Part III

	Author	*Work*	*Significance*
Chapter 11	John J. Mearsheimer	The Tragedy of Great Power Politics	Foundational text of the offensive structural realist school of thought. Criticizes defensive structural realism, highlighting that amongst great power expansion is not self-defeating. Highlights hegemony as the only true form of security major powers seek to attain, hence animating their competitive behaviour.
	Thomas P. Barnett	The Pentagon's New Map: War and Peace in the 21st Century	Offers a vision for post-Cold War American foreign policy, underpinned by a focus on structural rule setting and identifying globalization and interconnectedness as the central force of the post-Cold War international system.
Chapter 12	Robert D. Kaplan	The Revenge of Geography	The book provides a detailed overview of geopolitical considerations, focusing on why geopolitics continues to be relevant in the 21st Century. Important for demonstrating how natural geography can underpin state behaviour.
	Samuel S. Huntington	The Clash of Civilizations and the Remaking of World Order	The book proposes an ideational view on the motives of international politics, putting cultural and civilizational values at the centre of competitive interstate politics. While controversial for these notions, the book nevertheless provides important ideas on how identities and civilizational images can affect state interstate politics.
Chapter 13	M. Taylor Fravel	Power Shifts and Escalation: Explaining China's Use of Force in Territorial Disputes	The article provides an important model on how the timing of interstate wars can be explained through the dynamics of territorial disputes. It bridges the gap by moving beyond the fairly static concerns of higher level theories in order to identify triggers in day-to-day politics that can explain the actual timing of the commencement of war.

8 Systemic explanations

The grand theories of international relations often seek to attribute the act of war to some kind of systemic characteristics. An event with such high impact must have a grand cause. Realists approach the issue from the perspective of balance – or imbalance – of power and the hegemony seeking nature of survival-motivated states. Some other post-Cold War thinkers have also envisioned such systemic explanations for war. War is interpreted as a function of the international system as a whole, inherent to the nature of international politics and the distribution of power within the system. This section will focus on two key thinkers: John J. Mearsheimer, the foundational figure of offensive structural realism, and Thomas P. Barnett, who presented his core-gap thesis as a blueprint for post-Cold War American politics. Both thinkers have utilized a structure-focused approach where the international system is imagined as a complex interconnected structure, creating war due to the presence of structural forces.

Structural theories such as offensive structural realism create an important foundation for one's understanding for the occurrence of war, even if on their own they offer limited explanatory powers to make sense of specific conflicts. It is better to view systemic theories as a framework: They enable the space in which other causes of war can push combatants towards action. An imbalance in the distribution of power or hegemony seeking on its own does not inherently produce wars. But they amplify other considerations of insecurity and tangible strategic interests, such as geopolitical positioning, to push states to make the decision to go to war.

Balance of power and hegemony

The realist tradition most often associated with war is Mearsheimer's offensive structural realism. Unlike defensive structural realism, which considers war to be contrary to how states should behave most of the time, offensive structural realism, as the name suggests, puts an emphasis on the role the use of force plays in the international system. Mearsheimer provides a realist analytical framework for why a state would rationally choose to go to war, even if it is not defending the status quo.

The fundamental assumptions of offensive structural realism concerning the nature of the international system are largely identical to that of Waltz's defensive realism. Structural anarchy is viewed as central to the international system and Mearsheimer agrees that it creates strong incentives for security seeking and self-help behaviour. Mearsheimer identifies three key structural forces that animate state behaviour. First, anarchy is used in the same sense as previous structural realist works and refers to the idea that there is no central hierarchical authority that organizes the international system. Second, Mearsheimer departs from Waltz by arguing that states always possess offensive capabilities. The idea that there is a balance of offensive versus defensive weaponry has been important to Waltz's argument for status quo behaviour since it helped states make sense of the security dilemma.[1] In contrast, Mearsheimer argues that states always possess some offensive capabilities and that such a balance cannot be determined since the majority of weapon systems (e.g. a main battle tank) can be used both offensively and defensively.[2] This leads to the third structural force: Imperfect information. States can never be sure of other's intentions and it is prudent for them to assume the worst. According to Mearsheimer, an anarchic international system governed by imperfect information may push states with defensive intentions to nevertheless act aggressively.[3] It is this inevitable aggressive competition motivated by structural forces that puts the tragedy in the *Tragedy of Great Power Politics*. While Mearsheimer recognizes that states have the agency to defy structural forces, such a behaviour is severely punished by the system, often leaving states to be eliminated from the system altogether. In the end, while structural forces do not override agency, they do limit the potential number of successful strategies. This dooms states to engage in continuous competition.

Ultimately, offensive structural realism offers five key theoretical assumptions that underpin its interpretation of the international system, some of it a repeat of the aforementioned structural forces:[4]

1. The international system is anarchic.
2. States always possess offensive capabilities.
3. States can never be sure of each other's intentions.
4. Survival is the primary interest of the state.
5. States are rational actors.

While none of these factors alone would mean the adoption of the aggressive stance described by Mearsheimer, together they nevertheless lead to a system inhabited by revisionist great powers. They create three main patterns of behaviour that will underpin the actions of great powers: Fear, self-help, and power maximization. It is here where Mearsheimer most notably departs from Waltz's defensive structural realism. While Waltz argued for status quo behaviour, Mearsheimer argues for power maximization. The core thesis of offensive structural realism is that only by being the strongest in the system can a state be secure.

In the offensive realist system, every major power is inherently revisionist, motivated by these structural forces. They are constantly on the lookout for opportunities to expand or to alter the balance of power in their own favour. Why war would occur in such a system is rather obvious. The use of force represents an effective way to expand. In an anarchic system, the only break on the aggressive actions of the states is the balance of power. A war occurs if there is a breakdown in the balance of power, meaning that one state has gained substantial advantages over another. This allows the stronger state to use its might to secure advantages in the realist zero-sum international system. In this, Mearsheimer and Waltz offer similar views. The only major force putting constraints on states' expansive behaviour is the others' ability to resist. But while Waltz argues that this should motivate states to be satisfied with the status quo for the most part, Mearsheimer argues that they should actively work against balancing attempts by maximizing their power. In Mearsheimer's world balance is an inherently insecure position: Any sudden flux in the distribution of military power (e.g., a revolution in military affairs or a shift in alliances) can shatter the delicate balance, potentially putting the state's survival at risk.[5]

While balance of power provides the conditions that might allow the occurrence of war, Mearsheimer recognizes that state behaviour is motivated by ideational factors to a certain extent. Specifically, offensive realism recognizes the concept of "fear" as a central element of states' strategic calculations.[6] If a state is observing a negative shift in the balance of power, its insecurity rises and the state becomes fearful of the threats posed by certain states. The more afraid a state is, the more willing it is to engage in risky strategies to address the growing imbalance of power that is creating an existential threat. A good example of this is the politics of the Asia-Pacific. The United States often views China's strategies as unnecessarily risky, threatening a high chance of an armed confrontation for seemingly limited strategic gains. But from a Chinese perspective, there is a significant strategic fear of the United States, especially as the United States continues to direct its forces against the Chinese mainland, through the publicized development of the Air-Sea Battle concept or continued surveillance of China's nuclear submarine fleet. From the strategic perspective of the United States, these steps are viewed as normal day-to-day interactions between major powers. But for an insecure China, these are viewed as existentially threatening, necessitating a response, even if the risks are high. Harassing the USNS Impeccable might seem reckless for the United States. But from a Chinese perspective, it is costly signalling that China is deeply threatened by US naval surveillance in key waters.

If the balance of power is the primary determinant of the occurrence of war, then it is paramount to understand what power is. Mearsheimer argues that there are two key components to power: Latent and actual power.[7] Latent power represents a country's capacity to acquire actual power, but it is not directly involved in the balance of power. Latent power encompasses primarily wealth and population. Both of them are necessary to field a strong military.

Wealth is obvious. Modern weapon systems cost a lot. An F-22 Raptor fifth-generation air superiority fighter costs roughly a 150 million US dollar. While they represent the pinnacle of air power, only economically powerful states can afford to develop and maintain such weaponry. Population is a more complex issue. Automation should reduce the manpower needs of modern armies. While the US military is more powerful than ever, it uses a fraction of its Cold War manpower. However, with the increasing complexity of weapon system, the military needs to recruit highly capable individuals sought after by the private sector as well. Both Japan and China find it difficult to recruit the necessary manpower for their armed forces, despite the latter having a population of 1.6 billion people. A larger population would offer a better pool of candidates. While latent power represents the capacity to possess power, actual power is the foundation of the balance of power and states' ability to pursue interests in the international system. In its interpretation of power, Mearsheimer and offensive structural realism are seemingly more reductionist than, for example, Joseph Nye, who incorporates both hard power and soft power into his concept of smart power. However, this is partially a result of focus, rather than ignorance. As mentioned in the previous chapter, Waltz does not reject the importance of non-structural forces, but argues that to reveal the influence of structural forces we need to remove others from the equitation. In a similar fashion, offensive structural realism does not necessarily reject the existence of soft power but argues that hard power is the ultimate arbiter of interstate interactions and that soft power policies still need to be backed up by hard power. Thus, ultimately the balance of power is determined by the distribution of military capabilities.

Overall, we can see that in the structural realist tradition the occurrence of war is tied to the balance of power. It is an imbalance in power that allows for the use of force in the international system. At the same time, balancing is a strong deterrent against aggressive action. The primary distinction between offensive and defensive structural realism is their respective arguments of what to do with this situation. Defensive structural realism argues in favour of balancing and maintaining the status quo. Offensive structural realism in contrast argues for revisionist behaviour and power maximization. They both take the same systemic stimuli but come to diametrically opposite conclusions concerning their meaning.

Hegemony seeking

Adopting the logic of power maximization, the safest position for a great power is to achieve regional hegemony. Regional hegemony means that a state has gained such disproportionate amount of power that no other states in its home region could pose a direct territorial threat to it anymore. One example is the United States. There is no near-tier competitor in the Western Hemisphere that could challenge the United States. Its direct neighbours are Canada and Mexico, neither of which poses a substantial military challenge.

Similarly, Latin and South America contains no great power. Even countries like Brazil pale in comparison to the US military power. Furthermore, the United States has extended its sphere of influence on the entire region. It first used the Monroe Doctrine to remove any potential European challengers from the region. And today it acts as a security guarantor for the region, reducing the need of military build-up. Essentially the United States can be secure in the knowledge that none of its neighbours possesses the ability to pose a significant traditional security threat to it. The Soviet Union had occupied a similar position during the Cold War. Its massive military strength meant that no neighbour could have hoped to challenge it militarily. The only near-tier competitor was the United States. In contrast, the European continent has traditionally lacked a clear hegemon. With Germany, France, and the United Kingdom competing for regional dominance, none could achieve the security needed for hegemony. Britain has partially achieved hegemony: Its naval power was sufficient to insulate it from other continental powers and it allowed it to build a massive empire. But by the mid-20th century this position of seeming security was greatly challenged by continental powers' ability to reach across the English Channel.

The normative recommendation of offensive structural realism is that all major powers should seek a position of regional hegemony. And to a certain extent we observe this in the international system. The most commonly used example is China. The People's Republic of China (PRC) has been massively building up its military power as a foundation to its great power ambitions and Beijing has been seeking dominance in Northeast Asia. China clearly shares Mearsheimer's assumptions that the only position of security is being the strongest. But a similar logic can be observed in other countries. Japan has traditionally been viewed as an outlier when it comes to realist system dynamics. However, as discussed in the previous chapter, despite its economic focus, Japan has been seeking the construction of a hierarchical regional order with Japan on top. Japan clearly maintains an ambition of regional leadership. And under Prime Minister Abe, Japan has been moving in more realist directions, seeking a regional security leadership beyond its economic position in the region. While Japan does not necessarily seek to overcome China in military power, it is clear that the dynamic is becoming increasingly competitive, pushing Japan to re-emerge as an important actor in regional security.

There are two key patterns to the behaviour of hegemons: They compete with other hegemons and seek to prevent the emergence of new potential hegemons. The only existential threat to hegemons is the existence of other hegemons. As such, if more than one hegemon is present in the system, they engage in competitive behaviour in a mixture of balancing and attempting to overtake the other. A good example of such behaviour is the Cold War. The system contained two hegemons: The United States and the Soviet Union. While they could not be threatened by other countries, they viewed each other as existential threats. This led to a series of proxy conflicts, a competitive military build-up and decades of animosity. The main restraining force on their competition was the mutually assured destruction of nuclear weapons. But even short of a direct war, the United States and the USSR have sought to overtake each other in the balance of power. This led to a situation where everything was a question of balance of power and competition, including the space race. The first satellite, the first man in space, or the first man on the moon all became signifiers of the balance of power between the two hegemons. Both countries also tried to negate the nuclear stalemate. The United States especially has worked hard on missile shields and other weaponry that can eliminate the threat, and thus the restrictions, posed by an opponent's nuclear arsenal. This hegemonic competition created the most expensive arms race for a war that never came, eventually exhausting the Soviet economy, contributing to the collapse of the Eastern Block.

But hegemons are just as concerned about potential new hegemons as they are about each other. The only force that can restrict a hegemon is another one. Thus, the less hegemons are in the system, the more freedom and power each one possesses. The ideal position is to be the only hegemon, achieving global hegemony. The United States has enjoyed this position since the collapse of the Soviet Union, even if at times unsure of how to utilize its virtually unrestricted power. However, China is an aspiring hegemon. Achieving that status would disrupt the balance of power within the system, creating an equal competitor to the United States, restricting its ability to pursue interests. Thus, the United States has been viewing China as increasingly threatening and has been keen to hinder China's designs for regional hegemony. A major power can achieve regional hegemony by establishing dominance over its home region. The removal of direct immediate existential threats allows it to project power into the system and act as a hegemon. As such, the best way to hinder the emergence of hegemon is by preventing aspiring great powers from accomplishing such position of security. If the aspiring great power is insecure and preoccupied with home territorial defence, it won't be able to effectively project power abroad.

President Obama's "pivot to Asia" strategy was a manifestation of such strategic logic. The Obama Administration sought to shift away from the Bush Administration's focus on the Middle East and refocus the US strategic attention to the Asia-Pacific. This desire came to be known as the Administration pivoting to Asia. The pivot had several key components, including strategic rebalancing, doctrinal developments, and alliance building. As part of the plan, the Obama Administration has altered the distribution of American forces. Traditionally, the US Navy maintained a 50–50% split between the Atlantic and Pacific theatres. Under the Pivot, the US military sought to shift assets into the Pacific theatre, establishing a 60–40% split. Assets suitable to fight a great power war (air superiority fighters, surface combatants, strategic bombers, etc.) were moved into theatre, both to bases in Japan and to the US military hub of Guam.

The key challenge to US military power in the region has been China's anti-access/area-denial (A2/AD) defences. The PLA has been developing a strategy where it would deny the use of key access routes to opponents (primarily the United States) by deploying overwhelming force to achieve local superiority and to raise the costs of entry to unacceptable levels. This could allow China to win an engagement even if it is overall inferior to the opponent. The US military for a considerable time had taken access for granted and did not possess adequate strategies to counter China's A2/AD wall effectively. To address this, the Air-Sea Battle (ASB) concept or the Joint Concept for Access and Manoeuvre in the Global Commons (JAM-GC) was developed. The doctrine specifically seeks to dismantle A2/AD defences through the use of surgical force, including sea-borne airpower. This is achieved through the destruction of the enemy's command, control, communication, computer, intelligence, surveillance, and reconnaissance systems (C4ISR), shore-based missile and other weapon systems, and its command posts.

To maintain a forward presence, the Obama Administration sought to revitalize old alliances and form new ones. US military assets were moved into Australia, Malaysia and Singapore. US forces have returned to the Philippines. The United States even began a strategic rapprochement with Vietnam due to growing mutual concern over China. In addition, the Trans-Pacific Partnership (TPP) was envisioned to create a comprehensive framework for regional cooperation and US-led rule setting. China has been initially excluded from the process with the expectation that once in place, it would have no choice but to join on American terms.

The pivot officially made it clear that it is not targeting any specific country. It was marketed not as a hegemonic policy, but as a desire by the United States to focus on the Asia-Pacific due to the region's potential and bright future. However, it is hard to see the policy as anything else than an attempt to protect American hegemony in the face of a rising China. The policy is clearly designed to keep China insecure by bringing a military threat right into its backyard, and thus preventing it from establishing the secure position at home needed to become a hegemon. China has long proposed a division of the Pacific region, with the West overseen by China and the East remaining in the US sphere of influence. The United States has responded with a military build-up in the region. One can clearly observe the strategic logic emphasized by Mearsheimer. And China has definitely viewed the Pivot as an expression of hegemonic dynamics. Beijing has always argued that the Pivot is little more than the rebranding of the containment strategies used by the US against the Soviet Union that focused on local alliances to facilitate forward basing and doctrines specifically designed to counter cornerstones of the opponent's strategic planning. In essence, the Pivot was the US reasserting its hegemonic dominance by demonstrating to the Chinese that it has the power to dismantle Chinese defences and to project military power into mainland China if necessary. As such, the message was that China should act with care not to escalate to a military conflict, because its defensive strategies won't stop the US.

Keeping China insecure on purpose is a risky strategy. As Mearsheimer notes, fear is a good motivator for states to act assertively to escape insecurity.

Purposefully undermining Chinese security pushed China to become more assertive in its territorial disputes and to put a premium on a continued military build-up. But at the same time, it also limited China's ability to focus on theatres beyond its home region. Despite becoming the second largest military power, China has no presence comparable to the US or the Soviet Union outside the Asia-Pacific. While China has built up a military presence to secure its sea lines of communications (SLOC), its attention continues to be focused on the East and South China Seas and to resist perceived US containment efforts. And as long as the PRC remains insecure in its home region, it cannot claim to be a hegemon. Thus, the Pivot has achieved its strategic objectives, and did so without the need to engage in an actual armed confrontation. Shifting the regional balance of power was sufficient to deter any major Chinese attempts to overtake the United States in East Asia. In essence, by increasing the gap in the balance of power, the United States has avoided the need to militarily engage with China in the short to medium term. However, in the long term, it has increased strategic animosity, making it likely that China will be more assertive in establishing its own sphere of influence to escape insecurity. The logic proposed by Mearsheimer is clearly evident in this scenario.

The role of non-great powers

It is important to note that the main dynamics of offensive structural realism primarily apply to existing and aspiring great powers, hence the *Tragedy of Great Power Politics.* While the influences of structural anarchy are felt by all states, expecting minor powers to engage in power maximization and hegemony seeking would be unrealistic. Albania won't engage in a campaign of aggression to achieve regional dominance. This is not to say that these states are not concerned about the balance of power or act in a competitive and self-interested way. But their minor power status puts severe limitations on their ability to be competitive, including the fact that the system is ordered by the hegemons.

From an offensive structural realist perspective, minor powers are more of a component of the balance of power. Their behaviour is primarily understood in relation to the various hegemons through their choices of alliances. Waltz's defensive structural realism argued that states primarily engage in balancing as a response to structural anarchy. In contrast, Mearsheimer argues for a wider range of behaviours. He describes balancing and bandwagoning as the key patterns of behaviour for states.[8] Balancing is when states observe a shift in the distribution of power within the system and shift their posture to oppose it, deterring revisionist behaviour. While this would be normatively ideal behaviour to maintain the balance of power, it ignores states' self-interested and opportunistic nature. Sometimes a shift in the distribution of power offers an opportunity for smaller states to gain benefits by siding with the growing power. Bandwagoning refers to the behaviour when a state does not shift to counter a change in the distribution of power but rather moves towards the growing power. The hope is that by supporting the growing revisionist power it too may secure some gains in the ensuing reorganization of the system. This is crucial to explaining why expansion is not inherently self-defeating as Waltz would argue. Either way, through their associations, smaller powers shift the balance of power between hegemons by putting their own power, influence, and economic potential behind them. No great power can survive in isolation and they all tend to seek to carve out their own sphere of influence.

One can observe this in the contemporary competition between China and the United States. Both China and the United States seek to court smaller countries and to bring them into their own sphere of influence. In turn, these countries grant markets, basing rights, and resources that increase the power of the great powers. The United States has its established alliance system, a left-over from the Cold War. In Asia, the United States has mutual security arrangements with several countries. At the same time, Washington sought to unify the Asia-Pacific region under the TPP, which would ensure that trading is done under American-favoured rules. In contrast, China has its own regional designs: The strategic concepts of the Maritime Silk Road, the ambitious infrastructure integration of the Belt and Road Initiative (BRI), and the founding of the Asian Infrastructure Investment Bank (AIIB). These are each attempts by China to lure smaller powers into its own sphere of influence, away from the United States, in support of its own regional designs. The competition between the TPP and the BRI-AIIB in the Asia-Pacific has been the clearest example of hegemonic competition between the United States and China.

Overall, even if offensive structural realism is not primarily concerned about the behaviour of minor powers, it recognizes them as an important piece of the puzzle. They represent a value for great powers and are subject to the same competitive logic as any other component of the balance of power. Minor powers can exploit the competitive dynamics of the hegemons to get ahead by carefully managing whether to balance or bandwagon at any given moment. ASEAN is a key example for this.

Balance of power provides the basic foundation of why states go to war, but on its own it cannot fully capture the nuance of states' decision-making. The study of international relations puts an emphasis on the construction of analytical models to analyse state behaviour. In the construction of such models, one has to contend with two opposing forces. On the one hand, a model has to be sufficiently complex to capture reality and provide relatively accurate descriptions of state behaviour. On the other hand, the model has to be simple enough to identify the influence of key factors and to avoid being lost in detail or attributing undue importance to minor factors. Intelligence agencies for example have been studying the dietary habits of foreign leaders to gain insights into the behaviour of their states, although the importance of what Mao had for breakfast on the PRC's foreign policy is debatable. Waltz offers a good example of model building: Structural realism purposefully blackboxes a lot of factors, not because they are unimportant, but in order to reveal the importance of structural forces. In constructing one's own model, one has to follow similar principles. However, blackboxing too much will reduce the accuracy of the model. The existence of an imbalance of power is an important piece of the puzzle when discussing war as it creates a fundamental condition allowing a war to happen. But arguing that states simply watch the balance of power and have no other motivations animating their behaviour would be reductionist, leading to an overly simplified model with limited accuracy. As such, one needs to add nuance to the model by introducing additional factors. Many of these factors will be covered in the next two chapters.

The core-gap thesis and global norms

Realist analysis remains unconvinced about the importance of international norms and institutions. While such norms provide an important structure to the day-to-day interactions of the states, realists argue that in an anarchic system they encounter significant enforcement problems. In the absence of a higher authority, there are significant incentives to try to get out from complying with international norms when it can serve a state's interests. One good example of this would be how the United States was instrumental in setting up the International Court of Justice after World War II, only to deny its jurisdiction when it decided unfavourably in the *Nicaragua vs. United States* case. Peace and international order are not self-enforcing.[9] Rather some state or collection of states have to take this responsibility onto themselves and actively enforce the common ruleset of the system if it has to have any meaning.

Thomas Barnett offers a framework for how such a purposeful ordering of the international system might look like. In Barnett's view globalization is the key framework for the contemporary international system.[10] Globalization refers to interconnectedness and the acceptance of global norms and rules. Essentially, globalization refers to the creation of a somewhat global culture of what is acceptable and what is not for state behaviour. These rules are primarily based on Western ideals and are often shaped by the United States as the global hegemon in a unipolar system. Globalization in Barnett's analysis is viewed as a net positive. Adopting this global ruleset creates both a peaceful international system, through mechanics similar to economic interdependence theory, and prosperity for the individual countries.

The key issue for Barnett's analysis is that globalization does not have a global reach at the moment. Some countries have internalized globalization to a high degree, while other countries actively resist it.[11] If globalization is viewed as a net positive, then one has to ponder why anyone would resist its spread. Barnett lays the blame at the feet of local political elites.[12] Some regimes resist globalization to maintain their own power over the population, especially if they govern in a way incompatible with the global rulesets, for example, autocratic regimes. Similarly, some groups adopt resistance to globalization as a cornerstone of their insurgent movement. Whether one is talking about Iran, North Korea, the Taliban in Afghanistan, or Al-Shabaab in Somalia, the common trait is a desire for power over the local population, achieved by the purposeful isolation of said population from globalization and the rest of the world. One can observe an argumentation similar to Huntington (discussed in Chapter 9): Just as globalization is making a push to create a global culture, local forces arise to push back against it.

Based on this, one can create a dichotomic division of the world. On the one hand, one has the functioning core where globalization has taken roots strongly. This block is relatively peaceful as shared norms create a foundation for cooperation and for dispute to be resolved peacefully. On the other hand, one also has a non-integrating gap where globalization is low or non-existent. Hostile autocratic regimes rule these places, oppressive to their people and fearful of global influences undermining their powers. As Barnett puts it:[13]

> But look beyond globalization's frontier, and there you will find the failed states that command our attention, the rogue states that demand our vigilance, and the endemic conflicts that fuel the terror we now recognize as the dominant threat not just to America's future security but to globalization's continued advance.

It is the non-integrating gap that poses the threat to the functioning core. They undermine the global rules and often present a direct security threat to the core through terrorism or the financing of terrorism. The most severe conflicts are expected in the border regions of the core and the gap, where the struggle is most active to expand the reach of globalization.

The solution to the core-gap division and the threat posed by the gap is to actively shrink and eventually eliminate the gap altogether, and only then can the core achieve security. From a policy perspective, this can take many forms. Engagement and other influence-based tactics for spreading globalization are important. But the theory also has a more assertive side. If globalization is desired by most, and the main obstacle is local elites, then those elites have to be removed. This can be done through "peaceful evolution", introducing global norms to the population and waiting for the regime to crumble. The Soviet Union ended itself when it could not cope with the internal contradictions of its systems. But if this is not possible, then military force is required to remove threatening elements or obstacles to the spread of globalization. Barnett argues that the use of force in this system will focus on military operations other than war: Counterterrorism, counterinsurgency, and peacekeeping rather than traditional interstate wars.[14] But nevertheless, we see force being used for a political purpose, so for our purposes these acts would still largely count as "wars", based on the definition presented in Chapter 2. Essentially, the core's responsibility is to enforce the global rules and connect isolated countries, if necessary, against their will. This is so because countries outside these global rules undermine the stability of the system and encourage others to try to get out of obligations when convenient.

Barnett's core-gap thesis presents a very American view of global security. The United States does not face direct territorial threats being a regional hegemon. But even without that, the United States enjoys a position of largely unique geographical security. It is separated by any major powers with two oceans and it only borders two relatively minor powers (Mexico and Canada). The United States has not been territorially invaded in a significant way for centuries (much of its existence as a sovereign country). Rather, it is used to projecting power into distant regions based on its own immense economic resources. The United States does not share the history of invasion and territorial concerns of Europe or Asia. As such, the it tends to view its security in very different terms. China, for example, adopts a very territorial view of its security, supported by a long history of invasion. In contrast, the United States much more defines its security in terms of ideologies and compliance with rules. Washington is not interested in territorial invasions as much as forcing compliance with the US-led ruleset. Indeed, the last two attempts when the United States executed a territorial invasion turned out to be disastrous as decision-makers were not prepared for the challenges of territorial control and post-conflict reconstruction. They resorted to the use of military power to force a political change, but their reliance on invasion has thrown a wrench into the machinery. The difference between the 1991 use of force against Iraq and the 2003 invasion of Iraq is significant.

For great powers their influence within the system is a vital interest. They keep the system operating through their ability to organize states and set the rules of acceptable behaviour. The rules obviously are usually set up based on the preferences of the hegemon and designed in a way to benefit its own interests. For example, Wallerstein argues that the global free trade system is set up in such a way that it disproportionately benefits developed industrial countries while creating a constant flow of capital away from less developed ones.[15] This allows the hegemon to exercise power without constantly resorting to coercion or use of force. If states are allowed to defy these rules, the system begins to crumble and the power of the hegemon vanes. Importantly, this also creates an opening for rivals to present alternate norms and rulesets in favour of themselves, creating an ideational contest. A good example of this is the conflict between the Washington Consensus and the Beijing Consensus. Both are expressions of the values and norms of the two great powers within the system. The Washington Consensus emphasizes free trade and the interventionist promotion of Western values, while the Beijing consensus focuses on collectivist values, non-interference, and strict national sovereignty.

Hegemons and aspiring hegemons are powerful beyond the day-to-day considerations of traditional security. By their very nature they possess disproportionate military power within the system. Similarly, their interests are different from regular states. They focus much more on systemic contests and deal in global influence. Even for major powers it would seemingly make little sense to employ military force to defend values. Colin Gray specifically cautions against the use of force in a moralistic way or in an attempt to "do the right thing".[16] However, for hegemons these values and rules are the framework of the control they exercise over the system, and as such are a vital interest. When Barnett talks about expanding globalization, he is not talking about a decentralized global culture. Expanding globalization means expanding US hegemony over the parts of the world that have been resisting it for one reason or another. It is the completion of the US hegemonic project to achieve absolute domination within the system.

Aristotle called political science the "master science" based on a recognition that everything has a power dimension. Values, norms, and rules are not neutral. They always have an expression of power. Many interstate conflicts have been fought for ideational reasons, which offer a clear focal point to otherwise abstract interests, but most of these conflicts were at the end of the day about power: The power to shape an opponent and the system to one's own liking. One can normatively argue for a preference for free trade, human rights, or democracy as being desirable over oppression and isolation. However, one has to recognize that the promotion of these ideas is about power. Peaceful evolution might be viewed as cleaner than a war, but the results are the same: The pursuit of political goals. The United States promotes human rights and democracy in China because a democratic China is expected to be more compliant with global rules than one ruled by the Communist Party, thus posing less of threat to US influence due to shared values. The power of rule-setting and ideational forces is well illustrated by Japan. The United States managed to turn the potential regional hegemon into a stalwart US ally that largely supports the status quo. As such, sometimes the enforcement of rules and norms necessitates the use of force, even for a realist.

Notes

1 Waltz, K. (2010) *Theory of International Politics*. Waveland Press: Long Grove.
2 Mearsheimer, J. J. (2014) *Tragedy of Great Power Politics*. Norton: New York.
3 Ibid.
4 Ibid.
5 Ibid.
6 Ibid.
7 Ibid.
8 Ibid.
9 Gray, C. S. (2009) *Fighting Talk: Forty Maxims on War, Peace and Strategy*. Potomac Books: Dulles.
10 Barnett, T. P. M. (2004) *The Pentagon's New Map*. Penguin: New York.
11 Ibid.
12 Ibid.
13 Ibid, pp. 112.

14 Ibid.
15 Wallerstein, I. (2006) *World System Analysis: An Introduction*. Duke University Press: Durham.
16 Gray, C. S. (2009) *Fighting Talk: Forty Maxims on War, Peace and Strategy*. Potomac Books: Dulles.

9 National characteristics

The previous chapter has focused on the structural explanations of war, including the distribution of power within the system and the resulting balance of power, or lack thereof, which plays a role in creating the conditions enabling war. However, the explanations that countries would go to war for something as abstract as the balance of power or that the only motivation a country needs to go to war is a sufficient opening are rather unsatisfying. Survival is a good motivator, but it is often hard to conceptualize survival in the context of offensive military action. One has to reasonably wonder how going to war, especially as the aggressor, could ever serve one's survival interests. In a domestic context one is often conditioned to view hostility as inherently self-defeating and contrary to one's security. However, an anarchic international system is more akin to a post-zombie apocalypse world, where competition for survival is much different from an ordered domestic system. This chapter will focus on reviewing further considerations for going to war, which can provide a more tangible explanation of how a desire to survive can be nevertheless connected to the use of force, sometimes even its offensive use.

The chapter will focus on three major areas. First, it will discuss geopolitics and how the physical geography of the planet can provide incentives for war. States are often highly arbitrary creations out of synchronization with the inescapable natural geography of the planet. States inherently seek defensive positions provided by this physical geography, often creating the seeds of conflict. Second, the chapter will review the role of culture utilizing Samuel Huntington as its basis. The role of culture is a controversial topic in international politics, especially as one is rightfully hesitant to make judgements about different cultures. However, the formation of groups (including cultures) and the identities they create are an important factor of political behaviour, including the decision to go to war. Third, the chapter will examine the role of political organization in the decision to go to war, focusing on the thesis that political differences can amplify the likelihood of conflict as democracies are more likely to fight authoritarian states, while remain unlikely to fight other democracies.

Geopolitics

In his *Revenge of Geography* Kaplan proposes a radical notion: The physical geography of our planet is an important feature of our politics.[1] At first this seems self-evident. The idea that the physical configuration of the space we occupy affects our behaviour seems trivial. Yet contemporary political and strategic thinking often acts as if this is not the case. Technology has supposedly allowed us to conquer physical geography. Mountains, rivers, and deserts are irrelevant in the age of air power and cyber warfare. However, this overlooks just how much our political behaviour, including our strategic calculations that lead us to the use of force, is imbedded into this geopolitical context. Geography provides an inescapable framework for all human activities, including political contests. And sometimes, it is the very geography of the world that pushes states into armed confrontation.

Defensible natural boundaries

Kaplan makes the argument that strong natural boundaries exist on the geographical map.[2] Such natural boundaries consist of mountain ranges, rivers, or coastal lines that create naturally advantageous position to defend a country. These geographical features make it difficult for an enemy to advance and offer ideal defensive positions to halt an invasion. The strategic importance of terrain has been known at the very least since Sun Tzu, who argued about geographical positioning as one of the key components of successful strategic behaviour.[3] However, the political map, artificially constructed as it is, sometimes does not conform to the geographical realities of the topographical map. The attempts of states to reconcile these differences lead to contraction and expansion. And as any disruption of the status quo in a zero-sum environment, these expansions are often achieved through the use of military force.

To examine how such geopolitical considerations can motivate interstate conflicts, one should examine the competition between China and Japan concerning the Senkaku/Diaoyu Islands. The dispute holds a wide range of significance, but from the Chinese perspective one of the reasons is such geopolitical considerations.

Map 9.1 shows a basic overview of China's geography and security interests that underpinned its political behaviour leading up to the end of the Mao era. Starting from the central plains of China the initial phases of the Chinese Empire focus on the unification of the central rice-farming plains. This takes up the majority of early Chinese history. This is followed up by various projects of expansion and invasion (the Mongol invasion of China created the Han Dynasty). But eventually China reached the natural geographic borders that create a cohesive territorial unit. Leading up to the Qing Dynasty, Chinese history is a mixture of expansion and internal strife. External threats are limited during this period. In the east the littoral waters create a secure border with minimal threats posed by the occasional Japanese

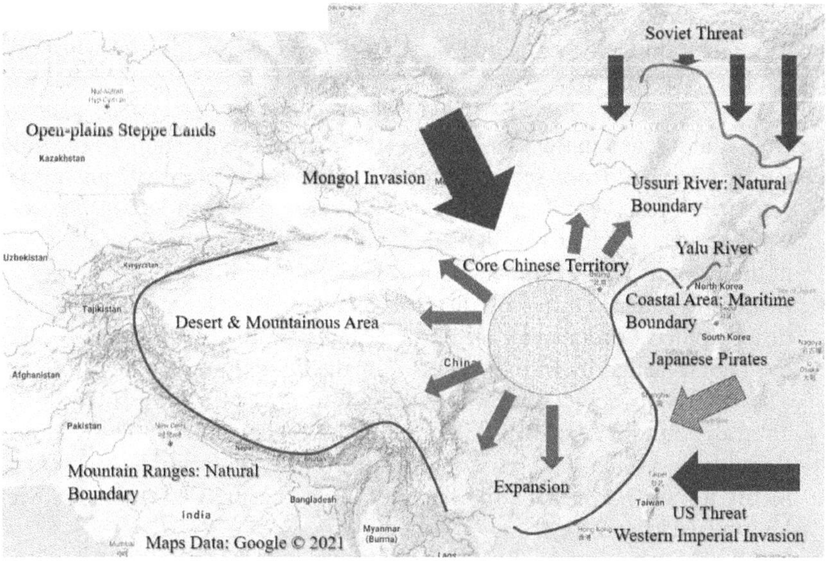

Map 9.1 China's geographical security

pirates. In the south and south-west, the mountainous regions similarly create a strong natural boundary. Beyond these boundaries China seeks to establish tributary relations instead of pursuing further territorial expansion, establishing indirect control through political and cultural influence. Chinese strategic planning predominantly focuses on the west and north-west, from where the nomadic people of the steppe lands present a continuous threat with limited natural boundaries.

Essentially, leading up to the 18th and 19th Centuries, China is largely geographically secure having unified all key territories under the Qing Dynasty. The Russian Empire is encroaching from the north, but the Ussuri and Yalu Rivers create yet another natural boundary, allowing a certain degree of security to the Chinese state. At this point the Chinese Empire is a continental power: China has very little interest in the maritime realm (considering it to be a natural barrier) and most political and strategic focus is dedicated to land-power and continental threats. China essentially maintains no seafaring navy and its defensive strategy focuses on riverine defence, for example, preventing access to strategic rivers that could lead an enemy from the coast to the heart of China's central plains. This is shattered by the arrival of Western imperial powers from the East, who utilizing naval power break down China's defences and begin the country's quasi-colonial domination, leading to the downfall of the Qing Empire. This maritime invasion creates the first breach in China's geographical security and will continue to underpin the country's growing insecurity that animates is behaviour until today.

If we jump ahead, Maoist China has fought four major land conflicts between the 1950s and 1970s. From the four, two were wars of compellence, one was a geopolitically motivated war that fits this reason for war, while one was a mix of the two:

- China intervened in the Korean War to prevent US expansion to the Yalu River. China did not seek territorial expansion of its own, but rather sought to preserve the security of its border with North Korea. The key Chinese strategic fear was that the United States would exploit a victory in the Korean War to launch an invasion against China.
- In 1962 China and India fought a border conflict as China sought to capture disputed border territory in the Himalayan region. The conflict has been complicated by India's support for the Dalai Lama.
- The 1969 Sino-Soviet War was about control over the Ussuri River, the natural boundary between China and the Soviet Union. The People's Republic of China (PRC) sought to gain control of several key islands to fortify its control over the river against an increasingly antagonist Soviet Union (see Map 9.2).
- The 1979 Sino-Vietnamese War was not about territorial expansion, but political manoeuvring. While China did launch an invasion beyond its natural boundaries, it did not seek to gain control of territory but to deliver a psychological blow to the Vietnamese in opposition to Hanoi's regional ambitions.[4]

On the East and South China Seas China is also motivated by geographical considerations. Following the death of Mao, Deng Xiaoping succeeded in the ensuing leadership struggle, capturing the reigns of the PRC for the moderates over the Maoist Gang of Four. This shift in leadership was followed up by a focus on economic development and opening. The focus of Deng's policies was to build up the Chinese economy through the creation of special economic zones. At the core of this effort was an export-oriented economic policy that was a significant departure from the inward-looking economic policy of Mao that emphasized self-reliance over integration into the world economy. Deng's reforms were highly successful and his economic model created the foundations of contemporary China and its economic might. However, these policies had unintended consequences. Traditionally China has always been an isolationist, inward looking power, with its core being the central rice-farming plains. Deng's economic reforms have shifted both the demographic and economic centres of gravity of the PRC. The new economic centres were created on the coast to feed into China's ports that linked the PRC to the world economy. Most of the newly created economic centres were all located in the coastal areas.

Essentially, while the coast was traditionally of little interest to China, under Deng's reforms they became China's economic centre of gravity. And with that economic development also came demographic changes. One can

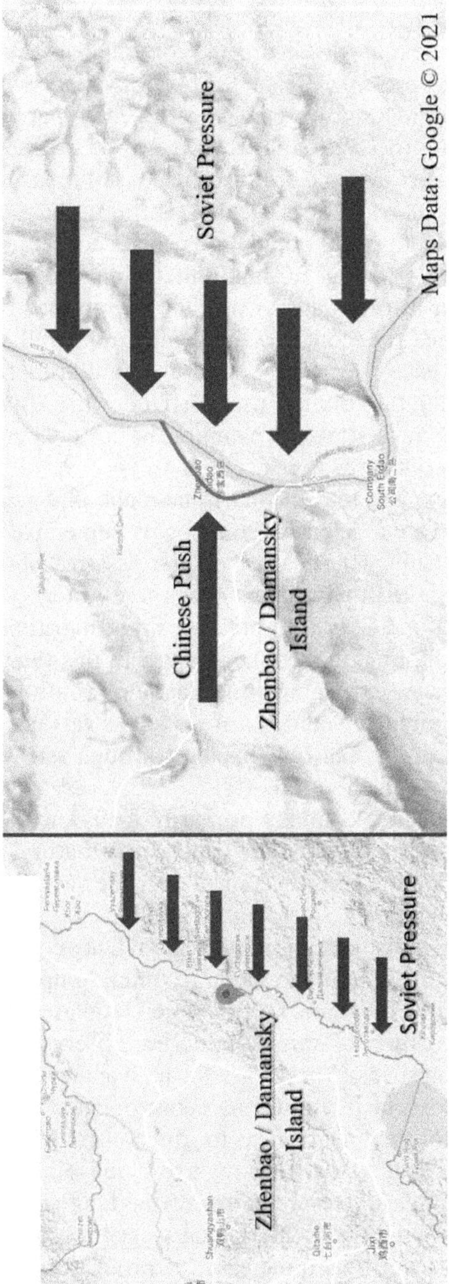

Map 9.2 Sino-Soviet border conflict

observe a significant migration from the central provinces towards the coast to provide the new industrial centres with a labour force, leading to the creation of major urban population centres in the coastal provinces. The meteoric growth of cities like Shanghai is well documented.

This shift has created a strategic conundrum for China. Traditionally, Chia has been largely geographically secure, the coastal areas creating a buffer zone isolating the central provinces from sea-born invaders. Mao's *People's War* doctrine specifically focused on engaging the enemy arriving from the sea on land, considering these coastal provinces to be expendable. This carried on a long history of relative neglect for the coastal areas in Imperial China. However, now these barrier zones became central to the prosperity of China. The industrial production and population centres concentrated in these areas make them essential. Yet, the PRC has limited ability to defend them due to a lack of control over littoral waters which are surrounded by potentially hostile powers. In essence, the economic and demographic shifts caused by Deng's economic reforms have negated the natural boundaries of China to a certain degree, rendering the country insecure. Under Liu Huaqing, the People's Liberation Army Navy's strategic planning began to focus on pushing the control of the PLA outwards at sea.

China needs to expand into the maritime realm and gain control over the littoral waters to create a defensive zone for its vulnerable coastal provinces. Unfortunately for China, the maritime space is already saturated by other powers who have staked their claims over these waters. China's traditional continental orientation and lack of presence in the maritime realm resulted in China having little footing on these key waters. In this vacuum Japan and the ASEAN states have largely claimed the East and South China Seas as their own, including claiming key land features such as various islands, islets, and reefs. China, at its point, can only expand through territorial disputes and confrontation with its neighbours.

The current territorial conflicts primarily focus on the control of strategic land features that will delimitate China's new natural boundary with its neighbours by facilitating China's control over the vast littoral waters of the East and South China Sea, shielding China's coastal provinces. China has been assertively pursuing its claims in the region. Once control has been gained over a strategic feature, China moves on to building up military facilities on these territories. While the islands offer limited land area, they offer strategic positions for the creation of missile batteries, observation posts, refuelling stations, and landing strips, if necessary through aggressive land reclamation. Each of these outposts facilitates China's control over the littoral waters. The ultimate objective for the time being is to gain control over the waters leading up to the first island chain. By attempting to exclude foreign warships from the area, including through reinterpretations of the United Nations Convention on the Laws of the Sea (UNCLOS), China seeks to restore a key component of its geographical security and increase the protection of its otherwise vulnerable coasts.

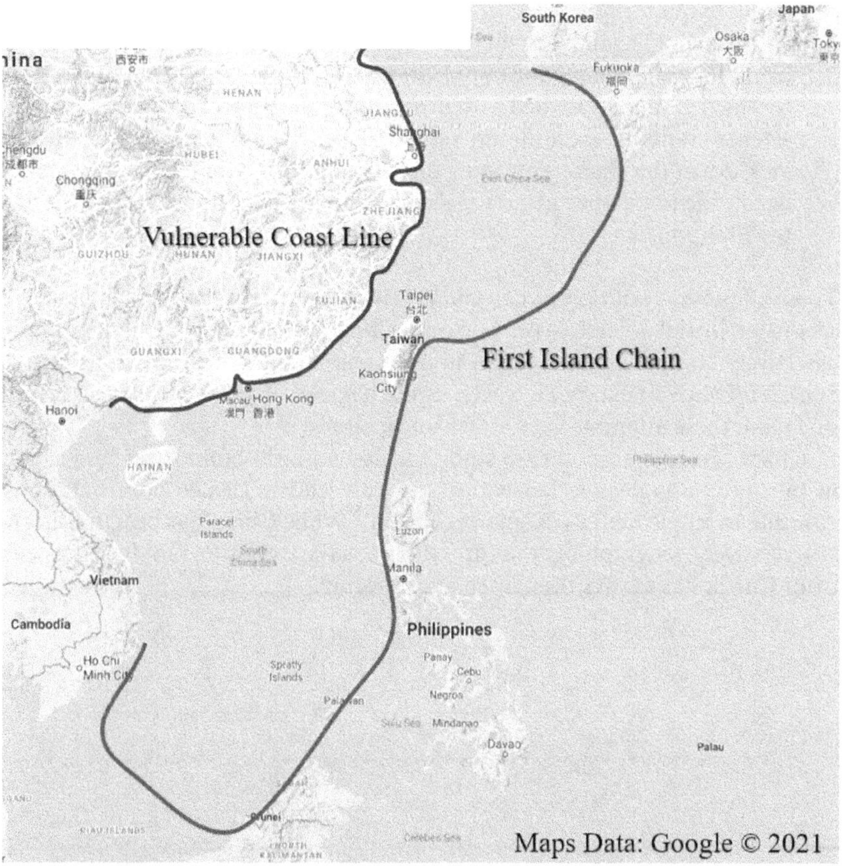

Map 9.3 Security of littoral waters

There are a number of confrontations China was involved in to secure these territories:

- In 1988, Chinese and Vietnamese forces engaged in a naval skirmish in the Spratly Islands concerning control over the Johnson South Reef. The engagement was a culmination of tensions concerning the ownership of reefs and other land features within the Spratly Islands group. The confrontation ended with a Chinese victory, allowing Beijing to establish control over parts of the Spratly Islands.
- In 1995, China employed military coercion to secure control over the Philippines-claimed Mischief Reef. China covertly constructed facilities on the reef. When this activity was uncovered by reconnaissance planes from the Philippines, the PLA-N stationed warships in the area to deter

a response. Due to the strong imbalance in power, the Philippines did not retaliate militarily.

- In 2012, China gained de facto control of the Scarborough Shoal. Vessels from the Philippines confronted Chinese fishermen operating in the area, to which China responded with dispatching law enforcement vessels. After a tense standoff, the Filipino vessels withdrew, and China proceeded to lock down the shoal, exercising effective control over it ever since. Once again, the imbalance of power deterred an actual military response from the Philippines.

These all present examples when China relied on force or the threat of force to secure territorial objectives on the South China Sea. Similarly, China has been assertively competing with Japan in an attempt to force concessions on the Senkaku/Diaoyu Islands. However, as the balance of power is more in favour of Japan, these attempts were so far unsuccessful.

China presents a good case study for how natural boundaries play a role in motivating aggressive behaviour and can lead to armed confrontations. Another example used by Kaplan is Russia.[5] While China has been traditionally relatively geographically secure, Russia is the opposite: On the European front Russia has always been open and insecure.

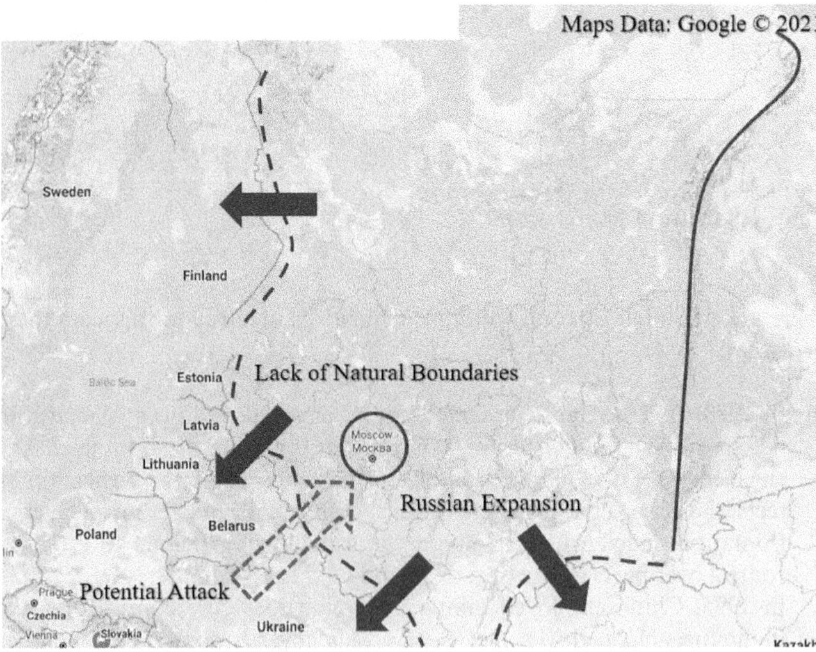

Map 9.4 Russia's geographical insecurity

This geographical insecurity has motivated centuries of Russian politics, seeking expansion in virtually every direction but especially into Eastern Europe, including Ukraine and the former Warsaw Pact countries. This contributed greatly to the Soviet Union's interest in securing territorial concession at the end of World War II and its aggressive expansion into Eastern Europe. It similarly contributed greatly to the ultimately failed Soviet invasion of Finland. And it contributes to the current ongoing tensions between Russia and North Atlantic Treaty Organization (NATO). NATO expansion into Eastern Europe was especially troubling to Russia. Russia was strongly against NATO's expansion into the former Warsaw Pact as it brings potentially hostile forces right into what Russia views as a buffer zone, and thus central to its geographical security. But if Russia was concerned about NATO expansion into countries like Hungary or Poland, Moscow was positively mortified by the possibility of Ukraine joining NATO under a leadership friendly to the West. The possibility of NATO expansion into Ukraine was viewed as such a severe threat that it triggered Russia's support for separatist movements in Ukraine and the ongoing "grey" military engagement. While it is tempting to dismiss the conflict as an autocrat being power hungry, the geopolitical underpinnings of the conflict present a more nuanced strategic picture. While in its contemporary form, most NATO members consider the organization as a vehicle for cooperation on security matters, it was formed as an anti-Soviet alliance and for Moscow it remains a force hostile to Russian interests. Analysis painting any expansion of NATO as neutral for Russia is wilfully ignorant at best. A similar trend can be observed in the case of the US missile shield in the region. Some showed confusion about why it received a strongly negative reaction from Moscow, seemingly ignoring the fact that it poses a direct threat to Russian second-strike capabilities, chipping away the foundations of Russian security, even if it is officially aimed at North Korea.

Fundamentally, we can see that when there is a disconnect between the political and geographical map, then states receive strong incentives for expansion, if necessary through military force. While this issue often overlaps with existing territorial disputes, that is not a precondition for it. Not every territorial dispute concerns natural boundaries and not every boundary issue is an existing territorial dispute. But the case for natural boundaries presents a strong additional factor beyond pure power politics that motivate expansionist behaviour. If anything, concerns over such boundaries are often dismissed as simply aggressive power politics. But without understanding the underlying factors correctly, one can have no hope of understanding the motivators of certain behaviours, which makes it supremely difficult to predict or effectively counter such behaviour.

Strategic territories

Moving beyond natural boundaries, there are territories that are not part of natural defensive lines, but nevertheless possess vital strategic importance to the state. Access to the seas is one of these. And even if one secures access to the seas, it has to be one that does not freeze over during the winter. This is the strategic conundrum Russia has encountered for much of its history. The Russian Northern Fleet is based in Severomorsks and patrols the arctic waters surrounding Russia in the north. These waters are prone to freezing over, limiting the fleet's mobility and patrolling. The fact that Russia lacks sufficient warm-water ports to field a large navy has haunted strategic planners for centuries. Russian expansion, including that of into Crimea, has been motivated by the desire to secure more suitable basing locations for Russian naval power.

A similarly thorny naval question is the need for nuclear-armed submarines (SSBN – nuclear ballistic missile armed submarine) to patrol securely to create a strong second-strike nuclear capability. The US solution to this problem was to patrol the deep open oceans where detection and interdiction would be supremely difficult. The Russian/Soviet navy did not have such a luxury. In every direction they are either cut off from the open oceans or the access is limited by cold weather and periodic freezing over of ports. The Black Sea is cut off by Istanbul and the Bosporus, while on its own it is not sufficiently deep to offer protection to SSBN. And even if that obstacle is removed, it is still only accessing the Mediterranean Sea. The Baltic Sea is surrounded by US allies and even if Russian submarines exit the area, access to the Atlantic Ocean is limited by the "GIUK gap" (Greenland – Iceland – United Kingdom) that offers a natural chokepoint ideal for anti-submarine warfare (ASW) operations. On the Pacific Ocean, the Japanese islands and the Korean Peninsula similarly limit access to the open oceans.

The Soviet solution to this problem was to create safe havens for their SSBN: Littoral waters where the Soviet Navy could dominate deterring ASW operations and ensuring the safety of its nuclear capabilities. This approach synergized well with the Soviet "new school" of naval strategy that emphasized the domination of littoral waters through a mixture of naval vessels, small patrol crafts, and shore-based assets. Today, China finds itself in a similar situation. Its own SSBN fleet similarly lacks access to the open oceans, limited by the first and second island chains. And Beijing too has opted for the Soviet strategy of safe havens. Its SSBN fleet based on Hainan Island seeks to patrol the South China Sea for cover. However, for them to avoid ASW threats, the PLA-N needs to be able to dominate these waters to deter interdiction attempts. At the moment, the United States is actively interfering with the security of China's nuclear submarine fleet by routinely conducting surveillance and mapping operations just off the coast of Hainan, relying on the provisions of the UNCLOS that provide the right of "innocent passage" to foreign navy vessels through a country's 200 nautical mile exclusive economic zone.

This led to the 2009 incident between China and the United States when Chinese vessels confronted the USNS Impeccable surveillance vessel. Chinese vessels, including a PLA-N frigate, approached the Impeccable at what is considered to be a dangerously close distance, forcing it to execute emergency manoeuvres to avoid a collision. The vessel also employed water cannons to deter the Chinese ships following it. The purpose of Chinese intervention was to force the Impeccable out of strategic waters where it was monitoring Chinese submarine activity. The conflict did not result in a military engagement, mostly because the United States, weary over a 2001 mid-air collision and the ensuing political crisis, withdrew the vessel. But a guided missile destroyer was dispatched by the United States to escort the Impeccable. It is not hard to see how such incidents could lead to an armed confrontation and how they signify the importance of strategic territories that present unique importance due to the geographic interests of one country or another.

Overall, one can easily see that the natural geography of the Earth can be a major contributing factor to competitive behaviour and can provide a basis for armed confrontations. Natural boundaries and strategically important territories underpin the security interests and defence planning of states. As such, they are highly contested: While one state seeks them to achieve security, others might counter these moves to secure power advantages or to keep an opponent insecure. This is one example where defensive strategic goals nevertheless might necessitate offensive actions to accomplish. In many of these instances, expansion is not motivated by a desire to gain power advantages or simply to pursue power maximization. Rather, often the goal is simply to escape deep-rooted insecurities. However, as Earth's finite space is already saturated, with every inch having been claimed by somebody, it nevertheless comes down to a zero-sum competition, often leading to armed confrontation between the claimant states.

Culture and clash of civilizations

The end of the Cold War and the collapse of the ideological dichotomy that underpinned it turned out to be a challenging time for International Relations theory. Realism as the major theoretical approach failed to foresee the collapse of the Soviet Union and as such suffered a blow to its credibility. Similarly, ideological considerations became moot point almost overnight with the rapid retreat of communism. International Relations theory needed new ideas to engage with the core mechanics of the international system, and a slew of them emerged. The post-Cold War period saw the revival of liberal economic-focused considerations, geopolitics (discussed above), and globalization (discussed in Chapter 8) as potential explanations for the workings of the international system. Culture also became a consideration, especially as constructivist ideational considerations became more mainstream. During this ongoing search, Samuel Huntington presented his thesis that it will be civilizational and cultural considerations that will underpin the key conflicts of the post-Cold War era, christening the theory the "clash of civilizations".

As we see with many International Relations theories, the division of the world into distinct areas is key to their understanding of the world. Realism focuses on the organization of major and great powers into competing alliance systems. Classical geopolitical theories focused on the distinction between continental and maritime powers and the heartland versus the rest of the world. Barnett focuses on the division of a core and gap based on the spread of globalization. Huntington follows the same logic when he divides the world into nine major civilizations: Western, Latin American, Islamic, African, Orthodox, Hindu, Buddhist, Japanese, and Sinic. These civilizations tie their respective countries together through shared cultural elements and customs. Huntington argues that the most important signifiers of a civilization are language and religion. The key argument of *Clash of Civilizations* is that with the demise of the ideological dichotomy of the Cold War, which has suppressed cultural differences, civilizational distinctions that contribute to the fragmentation of the world will lead to conflict.[6]

At the core of this division is the intrinsic mechanics of group identities. Any identity has two key components: An ingroup and an outgroup. The ingroup provides the definition of characteristics that determine who belongs to a certain group (or civilization in Huntington's case). In contrast, the outgroup is characterized by everyone else who does not belong to the group. But outgroup mechanics do not simply mean an absence of membership. Members of the outgroup are often characterized as the "other", something inherently different and alien to the ingroup. These memberships need to be naturally rigid to prevent people from switching to a group when it is beneficial but leaving it when it becomes a burden. In many societies such group memberships (e.g., kinship) act as a form of social insurance and provide the fundamental backbone of society.[7] The othering of the outgroup can create negative dynamics, such as distrust and animosity, or rhetoric focusing on shifting the blame to outgroup members for certain ills.

The ingroup–outgroup mechanics are reinforced by post-Cold War processes. Western civilization has emerged as the victor of the global struggle for hegemony. As such, Western norms and values underpin globalization with the ambition of creating a global culture (as discussed by Barnett in Chapter 8). As globalization continues to gain ground beyond the borders of the Western world, it is also facing a growing resistance. As the global culture becomes more homogenous, local cultures and identities are becoming more important. This is partially because of the mechanics of identity. An identity to be meaningful needs to have an outgroup. An identity with a universal or overwhelming membership tends to lack meaningfulness in pragmatic terms, since it cannot serve as a basis for distinction. This creates a resistance to the spread of Western norms and values and puts an emphasis on the existing civilizational differences. Highlighting how the other eight civilizations are different becomes to be perceived as valuable. The post-Cold War environment is consumed not only by a discussion on human rights, free markets, and individual liberty, but also by one on "Asian values" or the role of Islam in politics. Huntington argues that the post-Cold War environment will be primarily dominated by this civilization tension between Western civilization seeking global reach and other civilizations resisting these attempts.[8]

Huntington highlights that if we understand the international system to be based on such civilizational distinctions, then two key conflicts of great concern emerge. On the one hand, fault line conflicts represent confrontations on the borders of two civilizations. For example, India and Pakistan represent a conflict between a Hindu and an Islamic civilization. On the other hand, core state conflicts refer to the conflicts between the major states of a particular civilization. The United States attempts to spread global values and its engaging in military campaigns in the Middle East would represent such a conflict. In Huntington's view, these two conflicts present the clearest danger to the international system as ideational factors can most strongly exert their influence and push states towards a confrontation.[9]

The *Clash of Civilizations* is not without its critique. The notion that the contemporary international system would be motivated into large conflicts based on religion and culture remains a controversial notion, especially as cultural relativism remains a strong force. However, even if one does not go all the way to envision such a great clash within the international system, one cannot ignore the influence of culture on state's political calculations, including their willingness to go to war. Nationalism, as a political force, is based on the creation of "imagined communities" based on similar symbolic or ideational signifiers.[10] Nationalism has been an important motivating factor behind international conflicts, necessitating a critical look at the cultural and political-symbolic contexts of war. This section will link up with Chapter 3 at large, discussing the ideational context of war.

China presents a good example for how culture can affect the politics and threat perceptions of a country. The country has a long civilizational history, which then creates two trends that strongly contribute to the country's security interests and revisionist tendencies. China has some 5,000 years of history. For much of this history China has been a major or even great regional power, often enjoying a hegemonic position. The name "Middle Kingdom" has come from this position and China's traditional understanding of the world's political geography. China has traditionally understood the world much more in civilizational/cultural terms than the scientific topographic focus of European cartographers. Some classical Chinese maps envisioned the world more in terms of concentric circles: The innermost circle representing China as the centre of the "civilized" world, surrounded by a second circle of tributary states that benefitted from Chinese civilization, and then followed up by a third circle denoting "barbarians", societies who have not received the gift of Chinese civilization, including European states. In this organization it is clear as to why China bestows the name "Middle Kingdom" on itself.

This leads directly to one of the two key trends mentioned before. China views itself as a great civilization, with a role in regional leadership. It is viewed as an unjust regional order if China is denied such a position. However, the history of the country has meant that China has been largely denied such a position for a long time. The weakness of the Qing Dynasty in the face of Western imperialism, the struggles and internal divisions of Republican China, the Chinese Civil War, and Mao's communist isolationism all contributed to the fact that China has emerged on the world stage in the late 1970s and 1980s as a minor power. Due to its demographics, it had a potential to be a great power, but it was decidedly not. For much of its 19th and 20th century history, China has wielded little to no influence in international politics. This creates the psychological pattern of the Middle Kingdom Syndrome (MKS). At the centre of this is China's deep-seated desire to reclaim a position of regional and global political importance. From a Chinese perspective, it is unthinkable that a country with China's civilizational legacy and current economic and military might would not be in a central world-shaping position. One can observe a certain kinship with the United States: Just as the US politics is often shaped by American exceptionalism, China's own politics is shaped by its own sense of exceptionalism.[11] As Kissinger argued:

> China's splendid isolation nurtured a particular Chinese self-perception. Chinese elites grew accustomed to the notion that China was unique – not just "great civilization" among others, but civilization itself.
>
> As the Chinese saw it, a host of lesser states that imbibed Chinese culture and paid tribute to China's greatness constituted the natural order of the universe.[12]

The desire to reclaim its position as a great power is also incorporated into China's nationalist/patriotic narrative. The narrative received prominence as the Chinese Communist Party's (CCP) legitimacy has declined after the economic opening due to the diminishing of its Marxist–Leninist ideological roots. As such, the CCP shifted its focus towards patriotism to solidify their fractured legitimacy. A component of this was emphasizing external threats to the country and the Party's efforts and achievements to counter them and right historical wrongs. The CCP is set up as the vanguard of China's efforts to restore the country to its rightful place under the sun, the Communist Party being uniquely suited to undertake this task. These narrative points have been especially prominent in the politics of Xi Jinping who had departed from traditional CCP wisdom of "hiding strengths and biding time" to openly display China's great power capabilities and ambitions.

This narrative works because of another trend: A deep sense of historical shame. Callahan argues that China's political identity is dichotomous. There is a great sense of optimism about the future, but also deep-seated anxiety about achieving it. Civilizational achievements offer a great sense of pride, but history is also a source of shame. Callahan describes this dualistic attitude as "pessoptimism".[13] The main source of this dichotomy is what is known as the "Century of Humiliation". The Century of Humiliation is part of China's patriotic narrative and it describes the roughly 100 years from the 1st Opium

War (1839–1842) to the formation of the PRC (1949). This is a period of significant weakness that saw the downfall and quasi-colonial subjugation of the Middle Kingdom. It is also the period during which China has suffered the brutal war against Japan, a country it traditionally considered a recipient of its own civilizational gifts. Even in the Century of Humiliation there is a shred of exceptionalism, viewing Chinese history as exceptionally tragic, despite the fact that the late 19th and early 20th centuries were miserable for a large part of the world. Hungary, for example, has a history that parallels China's. After a successful national revival movement, it suffered a defeat in its rebellion against the Habsburg Empire in 1849. It was dragged into World War I, leading to significant territorial losses. After 1918, the country suffered a period of instability and infighting, before being dragged into World War II, leading to the Soviet occupation of the country. Hungary developed a similar narrative of historical tragedy, albeit lacking the focus on humiliation present in China's Century of Humiliation narrative..

These are important components of China's contemporary identity and penetrate into its culture. They are relevant to the discussion at hand because they also underpin China's strategic behaviour, including its competitive stance against other regional states. Due to these ideational components, China is highly distrustful of foreign powers and puts an enhanced emphasis on its own security. It views the international system as hostile and threatening, necessitating strong security seeking behaviour. Essentially, these cultural components push China to be the ideal offensive structural realist state: Scared and seeking power maximization due to viewing strength as the only reliable position of security.

This cultural focus on historical loss and recovery also motivates some of the country's territorially revisionist policies. Territories such as the Spratly Islands or Senkaku/Diaoyu Islands are viewed as "lost territories" – territories lost during the Century of Humiliation – and as such they gain high political-symbolic importance beyond their immediate strategic and economic value. Beijing cannot compromise on these territories without going against the political narrative they promote to solidify the legitimacy of the CCP. This was especially obvious in 2009. Then Secretary of State Clinton announced to an increasingly concerned ASEAN that China has declared the South China Sea a "core interest", the highest level of Chinese interests, similar to Tibet or Taiwan, which is non-negotiable and subject to use of force. This coincided with the launch of what became the Obama Administration's "pivot to Asia". China made no such declaration. But it could not come out and deny that the South China Sea was a core interest, even though it would have limited the political fallout, without going against its own political narrative. Thus, China's culture of nationalism contributed to the escalation of tensions in the region. As one observer put it:

> It is not Chinese policy to declare the South China Sea as core interest. But the problem is that a public denial will be some sort of chicken action

> on the part of Chinese leaders. So the government also doesn't want to inflame the Chinese people.[14]

As one can see, culture, and the history connected to it, is an important part of the puzzle. It can magnify insecurity and dictate strategic objectives. It is a fundamental component of the cognitive framework of decision-makers and one can only ignore it at their own peril. Mearsheimer recognizes the importance of fear in international politics and how it motivates risky or aggressive behaviour. Yet, he defines fear in purely balance of power terms. This is of course not unexpected for structural realist analysis, but it is a reductionist approach nevertheless. The balance of power itself is intangible, only existing as a perception filtered through the state machinery. But the meaning of the balance and how much threatening it is perceived are strongly influenced by the culture of the country that then affects the leader's cognitive framework. In the case of China, its cultural context makes it more scared than the pure balance of power and geopolitical considerations would suggest, making it more predisposed to deploy force to achieve interests deemed vital for its own security.

The acceptability of use of force in general is affected by the culture of a particular nation. In Chinese strategic culture there is a high acceptability for use of force. Kissinger denotes that China views use of force more in terms of psychological negotiation, akin to Schelling's diplomacy of violence.[15] The country's military doctrine also focuses on "active defence" which puts an emphasis on the pre-emptive deployment of force to counter potential threats, based on the strategic logic that China will not sit and wait to receive a first hit if it is determined that a conflict is unavoidable. Similarly, the United States has demonstrated a willingness to employ military power. It maintains a significant military presence across the globe, and it has intervened militarily in several conflicts since the end of the Cold War. It is also highly comfortable with utilizing drone strikes and to carry out cover operations on the soil of other countries, often without the permission of the local government. In contrast, Japan is far more reluctant to resort to military might due to its own anti-militarist culture. Even in UN authorized peacekeeping missions, the deployment of Japanese troops to foreign soil has been controversial. Its own strategic doctrines emphasize war as an absolute last resort, and, prior to the reforms of Prime Minister (PM) Abe, only in a case where Japanese home territory has been directly targeted. And PM Abe's legislative efforts to loosen restrictions on the deployment of Japanese military power remain highly debated, often condemned as "war legislation".

Culture provides an expression of the cognitive framework of a society. They express what is right and what is wrong. It's a key ideational component that cannot be ignored when analysing the possibility of war. While it is important not to be deterministic or to be trapped in arguments of cultural superiority, the culture of a particular society can show whether it is predisposed to be more or less willing to employ military power and what interests does it define as acceptable for the use of force. One does not have to envision a clash of civilizations in the manner of Huntington. But one cannot ignore that culture presents an important point of division in the existing international system. Its influences penetrate all facets of politics: What does a nation desire, who it views as friends and enemies, how it seeks to achieve its interests, how many strategic threats does it perceive, how severe its level of fear is, and so on and so forth. Without understanding how a country thinks, one cannot understand their potential for war. And without culture and identity, one cannot understand the cognitive processes of any state. Sometimes symbols of political and cultural significance are worth dying for. Or at the very least, they worth dispatching military force so that others would die for it.

Democratic peace and authoritarian war

The argument that democracies don't fight each other is not a new one. Russett has highlighted that democracies rarely fight because (a) they have developed other means of conflict resolution and (b) their democratic identities predispose them to view war as unacceptable against other democratic states.[16] This argument links up with Mueller (discussed in Chapter 7) on the evolution of the international system. However, Russet notes several caveats in this argument. First, this is not an absolute prohibition on war, simply that democracies fight each other less. Second, this does not mean that democracies are more peaceful overall.[17] Against authoritarian regimes (or other non-democratic political systems) democracies are just as likely, if not more so, to use force as any other state. If one accepts Russet's thesis, then the political organization of states is a key factor in determining states' potential to go to war against each other. This is especially important in an Asian context. While Europe, for example, is largely occupied by democratic states, Asia is home to some of the largest and most powerful non-democratic regimes, such as China, and non-democratic values continue to be prevalent in the region. Some of the most contentious conflicts in the region are between democratic and non-democratic states, for example, China versus Japan, further raising the risks of war.

The central ideas of democratic peace theory are intrinsically tied to the nature of democratic institutions. A democratic government relies on consent to a much higher degree than a non-democratic one, including the consent of those who will be sent off to die in the ensuing armed confrontation. We can refer to the ideas of De Mesquite and Smith. Any society has three key groups of people: The nominal selectorate, the real selectorate, and the winning coalition. The nominal selectorate (interchangeables) are the people who potentially participate in the selection of the leadership process. The real selectorate (influentials) are those whose support is needed for governance. And the winning coalition (essentials) is the group that is absolutely essential for victory. This has profound implications for governance. Autocracies have small influential and essentials, drawn from large interchangeables. In contrast, democracies have large influential and essentials, the influential largely matching the interchangeables.[18] In such a society, a broad-based consensus needs to be built in order to go to war, which can be difficult. This can be especially so in the case of a major or total war where large segments of the influentials would face death or depravation. While this can be justified in case of a war of resistance, it is much harder to justify for a war of aggression in the pursuit of an abstract national interest, especially against another democratic state. And once at war, it is much harder to sustain it as casualties and economic damage mount, as the United States has found out in Vietnam. Autocracies have less problems with these as most with actual political power would not see active combat or suffer many of the other consequences of war. This allows them to sometimes pursue war as a foreign policy tool more freely than a democratic country. The internal structure of governance thus exerts a significant influence on the state's capacity for war.

The idea of democracy affecting a country's potential for war connects up with diversionary war theory (discussed in Chapter 10) in one major way: The fusion of domestic and foreign policy considerations. While war is inherently international, the form of government is equally inherently domestic, falling under the sovereignty of the state. This raises the question of how and how much a state should interfere in the internal affairs of another. Democracy is often viewed as universally desirable. But at the same time, there is plenty of evidence for the appeal of strongmen governments and societies' willingness to forgo many aspects of participatory democracy in favour of stability and security. Furthermore, sovereignty as a theoretical absolute continues to be an important component of the international system. Thus, from a foreign policy perspective it is especially challenging to put its tenant into action.

The theory in a way is similar to Barnett's "new map" (discussed in Chapter 9) because of its focus on a dichotomic division of the world into "good and bad guys", rule followers and rule breakers. Rule breakers, who do not conform to the established expectation of democratic governance, are viewed as a threat to the rest of the system, necessitating countering action and at the very least defensive strategic planning. This "us versus them" division then creates the root of conflict as both sides begin fearing and distrusting the other. It is not just that democratic countries fear the potential of autocratic states for war, but that autocratic states fear democratic states' intervention to topple their existing regimes.

This is expressed in many political discourses (e.g., in the case of China). The idea of "peaceful evolution" is a major concern for the CCP. This originates in the post-Cold War environment. The United States and its allies hoped that by engaging China and exposing its population to external norms and values, including democratic governance, they can trigger an internal shift that would result in the downfall of the Party, without the need to fight China. Obviously, the Party did not appreciate these ambitions. Internally, this contributed to China's growing focus on surveillance and nationalism. Externally, this reinforced China's distrust and hostility towards foreign powers, especially as this confirmed the existing narrative on the prevalence of foreign threats in a country where regime survival and state survival are considered one and the same.

Thus, while democratic peace may have the best intentions of reducing interstate conflict, its focus on essentially Western rule-setting promotes a division that underpins the threat perceptions of some powers. Here one can connect back to Huntington. Democracy might be quantitatively a significant factor in reducing interstate conflict, if spread globally. But it is arguably a Western civilizational value, with other cultures being more accepting of communalistic or autocratic values. And autocratic regimes are both influenced by these cultural values in their threat perceptions and capitalize on the existence of these values among the greater public to facilitate their own political survival.

Ultimately, division is a source of conflict. If there is an "us" and a "them", it will lead to disagreements and competition, and sometimes to an armed confrontation. Once again, the role of ideational factors should not be underestimated in a state's capacity for war. To refer back to Waltz, it is possible that the pursuit of peace would require more bloodshed than an anarchic and confrontational international system would have produced otherwise.

Notes

1 Kaplan, R. D. (2013) *The Revenge of Geography*. Random House: New York.
2 Ibid.
3 Sawyer. R. D. (2007) *The Seven Military Classics of China*. Basic Books: New York.
4 Kissinger, H. (2012) *On China*. Penguin: New York.
5 Kaplan, R. D. (2013) *The Revenge of Geography*. Random House: New York.
6 Huntington, S. P. (1996) *The Clash of Civilizations.* Simon & Schuster: New York.
7 Collier, P. (2010) *Wars, Guns and Votes: Democracy in Dangerous Places*. HarperCollins: New York.
8 Huntington, S. P. (1996) *The Clash of Civilizations.* Simon & Schuster: New York.
9 Ibid.
10 Anderson, B. (2006) *Imagined Communities: Reflections on the Origins of Nationalism.* Verso: London.
11 Callahan, W. A. (2012) "Sino-Speak: Chinese Exceptionalism and the Politics of History". *The Journal of Asian Studies* 71(1), pp. 33–55.
12 Kissinger, H. (2012) *On China*. Penguin: New York, pp. 10.
13 Callahan, W. A. (2010) *China: The Pessoptimist Nation*. Oxford University Press: Oxford.
14 Wong, E. (2011, March 30). China Hedges Over Whether South China Sea Is a "Core Interest" Worth War. Retrieved 18 May 2011, from *New York Times*: www.nytimes.com/2011/03/31/world/asia/31beijing.html?_r=4&ref=global-home.
15 Kissinger, H. (2012) *On China*. Penguin: New York.
16 Russet, B. (1994) *Grasping Democratic Peace*. Princeton University Press: New Jersey.
17 Ibid.
18 De Mesquite, B. B. and Smith, A. (2011) *The Dictator's Handbook*. Public Affairs: New York.

10 Practical considerations

The two preceding chapters focused on larger considerations such as structural forces, or the fundamental nature of states and the planet. This chapter adopts a decidedly more limited focus. Not all wars are grand contests of survival or clashes of civilizations. Some, if not most, wars are limited contests over political disagreements. This chapter focuses on reviewing the role of the smaller practical considerations for the occurrence of war. While these factors can be, and are often, tied to larger considerations, they play a unique role in explaining not only why a country went to war but also why it went to war at a particular time. One of the key shortcomings of the aforementioned approaches is their ability to explain the general conditions that lead to the outbreak of a war, but struggling to offer a persuasive explanation of why a country went to war in one year rather than another due to their considerations being fairly static. This chapter looks at escalation of territorial disputes and the possibility of diversionary wars as specific triggers that can capitalize on larger considerations mentioned above to throw the iron dice of war.

Escalation of territorial disputes

Territorial disputes are arguably some of the international disputes most prone to leading to an international conflict. Territory has been important to explaining the occurrence of war.[1] There are a number of reasons for this. Unlike resource control, where joint development offers a good way to compromise, under our current largely Westphalian understanding of the international system sovereignty over a territory is indivisible. Territory is inherently zero-sum: Only one state can exercise control over a territory and any gain must come at the expense of another state as all of the surface of the planet is claimed. This makes a compromise, short of accepting only partial territory, quite difficult. This is not to say that resolving territorial dispute peacefully is impossible. There is a long history for resolving territorial conflicts through bilateral negotiations (China has resolved many of its territorial disputes in this way, including the Ussuri River island disputes with Russia), non-binding third-party intervention (Papal mediation played an important role in resolving territorial disagreements between Chile and

Argentina), or binding arbitration or adjudication (Malaysia and Singapore resolved the Pulau Batu Puteh dispute in this way). However, in many cases such options prove to be ineffective, especially if the territorial dispute possesses an entrenched political-symbolic significance or is of geopolitical or security importance (discussed above). In these instances, escalations are common, paving the way for a military confrontation between the claimant states. This is so even if the dispute has been dormant for a long period of time, which raises the question of how escalatory mechanics work and contribute to the breakdown of the status quo.

The foundations of territorial disputes are not hard to understand: Territory represents value. This value can manifest in many different ways. The most obvious ones are economic and strategic advantages one can secure by possessing that territory. Islands can offer vast economic benefits by securing exclusive economic zones under the United Nations Convention on the Laws of the Sea (UNCLOS). But even beyond that, some territories are of cultural significance. Thailand and Cambodia have been in a bitter dispute over ownership of the Preah Vihear temple complex. Mediaeval crusades were motivated by the cultural and religious significance of the Holy Land to Christian European nations. Some territories get incorporated into the nationalist narratives of countries, serving as a rallying point for political organization. China considers the Senkaku/Diaoyu Islands to be "lost territories", incorporating it into its national humiliation narrative and the mythology of Chinese resistance against Japanese aggression. Some territory is valued because of being inhabited by groups sharing an ethnic background or is part of a common history. Since territory provides the spatial expression of human existence, possession of territory is always valued. This can be compounded by other factors, such as the level of democracy in a certain country, internal political processes necessitating diversionary actions, or an imbalance of power between the claimant states. However, Fravel points out that these explanations, while important considerations, fail to capture the essence of escalation as they tend to be fairly static and unchanging.[2] They create the foundations that lead to an armed confrontation, but on their own cannot explain the timing of armed confrontations or why disputes tend to remain dormant for considerable periods of time. There simply must be other factors that lead to critical escalation and the onset of war.

Rasler and Thompson argue that territorial disputes are especially prone to violent escalation as they create ideal conditions for rivalries to be entrenched between states.[3] Rivalries are ideational concepts that represent that animosity between states goes beyond a simple clash of interests and acquires an identity of their own. As discussed in Chapter 7 of this book, Wendt highlighted that ideational considerations, including state identities, offer an important dimension to interstate politics. As Rasler and Thompson argue:

> Territorial disputes between rivals act as lightning rods for all the psychological baggage and mistrust associated with protracted antagonisms.

> However genuine and intense the conflict over territorial possession, such disputes become convenient vehicles for combating the external enemy.
>
> It works the other way around as well. Without something concrete as a territorial dispute, rivalries may seem too abstract to large portions of the publics involved. [4]

Adopting a constructivist perspective, the idea that rivalries, built up through territorial disputes and other negative interactions, would create a negative association in security makes sense. And in turn, the competitive dynamic diminishes incentives for a compromise while instilling a sense of insecurity into the rivals. One can certainly observe such a rivalry in the case of China and Japan. China considers Japan to be one of the major obstacles to realizing its own ambitions of great power status. The animosity between the two countries however goes beyond a mere strategic competition. China sometimes engages in diplomatic skirmishes that seem somewhat petty. One good example of this is the "Voldemort Wars", an incident during which both countries insisted that the other is the manifestation of the main antagonist of the "Harry Potter" book series. It was the Japanese ambassador to the United Kingdom who first made the comparison. However, China could not let the incident go and retaliated with its own comparison, creating one of the more amusing clashes of contemporary diplomatic history.

However, while rivalry is an important consideration for the escalation of conflicts (rivalling states are more likely to be confrontational), it too is a fairly static component. As any ideational components, rivalries do not come into existence overnight, nor do they diminish rapidly, lingering long after the original causes might have subsided. Japanese Prime Minister Koizumi's visits to the controversial Yasukuni Shrine have contributed greatly to Sino-Japanese rivalry. However, China did not de-escalate significantly after PM Koizumi's successor Prime Minister Noda adopted a less provocative stance, bringing Pacific economic cooperation as the focus of his agenda. Rivalry in this sense is more similar to the value of territory, a fairly constant factor. It can create the conditions for escalation and war but offers limited insights into their precise timing.

Fravel in contrast argues that it is fluctuations in bargaining power within a dispute that motivate escalatory behaviour and can trigger an armed confrontation.[5] Bargaining power in this context generally refers to the strength of a country's claim relative to the other claimants. The stronger the claim, the more likely the dispute can be resolved with a compromise favourable to the country. In contrast, if the claim is weak, a country can expect a bad compromise or even not getting anything at all. Fravel defines bargaining power based on two key components: Amount of land held and military power that can be projected into the disputed area.[6] The combination of these factors determines the relative balance of power within the dispute. Both factors should be fairly obvious. The more land a state possesses, the more it controls the situation, at least through *de facto* control. Similarly, the more military power a state possesses that it can project into the dispute, the more options it has to exert direct influence if negotiations fail. In contrast, if a state possesses little of either or both, then its position in the dispute is severely weakened.

Naturally, these factors are subject to fluctuation, especially when it comes to the military balance between the two countries. If a state is experiencing a significant negative shift in its bargaining power, it needs to take corrective action to address the imbalance. At its core, this mechanic is similar to that of the basic logic of realist power politics focusing on the balance of power. States seek to both balance and to seek out advantages. But a negative shift is inherently threatening to their position necessitating corrective action. In a territorial dispute a negative shift in bargaining power is inherently threatening because it means that the state would likely have to accept a worse deal as a settlement in the future or won't be able to prevent a military annexation of the territory. Hence, a response is needed to both prevent a further negative shift and to deter an opponent. This usually manifests in the form of pursuing escalatory policies, seeking to raise the perceived costs of increasing bargaining power for the other state.

A number of things can contribute to a sense of a negative shift. The most obvious ones are the ones that directly alter one of the two components of bargaining power. A military build-up in the disputed region or the actual seizure of land represents concrete signs of a negative shift. Both can be observed to play out on the South China Sea. The 1974 Paracel Islands confrontation offers a good example of this. People's Republic of China (PRC) and South Vietnam shared the Paracel Islands group on the South China Sea, albeit both claiming sovereignty over the entirety of the group. The island group can be divided into two major parts: The Amphitrite Group in the northeast, originally controlled by the PRC, and the Crescent Group controlled by South Vietnam. The prologue of the dispute is rather similar to that currently observed on the East China Sea. On 11 January 1974, Beijing issued a formal statement reaffirming indisputable Chinese sovereignty over the islands. South Vietnam condemned the declaration reaffirming its own sovereignty over the Paracel Islands.

The next phase of the events began with two Chinese fishing vessels entering into Vietnam-controlled waters on 15 January. Vietnamese warships patrolling the area opened fire on them with the intention of driving them out of disputed waters. Vietnamese personnel landed on Robert and Money Islands, in the Crescent Group, to remove PRC flags supposedly planted by the crew of the fishing vessels.

South Vietnam followed up the removal of the flags by landing troops on Robert and Money Islands again on 17 January. The landings were observed by the PLA-N and Beijing ordered two Hainan-class fast attack crafts to return to Woody Island, in the Amphitrite Group, to pick up Chinese naval militia troops and return to the area. In the meantime, the South Vietnamese navy once again encountered the two fishing vessels from the 15th and proceeded to confront them until the PLA-N issued a warning upon which Vietnamese ships abandoned the pursuit. However, the following day, Vietnamese ships confronted the fishing vessels for a third time, this time one of the destroyers ramming one of the fishing vessels, according to the official statement concerning the accident. At night both navies dispatched reinforcements to the region. On the eve of battle, both naval forces had a core of four ships: The RVN commanded three frigates and one corvette, while the PLA-N had four corvettes at its disposal. The initial balance of forces favoured the RVN.

The combat phase of the confrontation began on 19 January. RVN troops began landing on Duncan and Palm Islands, in the Crescent Group, when they were confronted by PLA-N troops and Chinese naval militia. Accounts conflict on who fired the first shot. Vietnamese sources claim that their troops were attacked by the Chinese troops on the islands, while Beijing claims that Vietnamese troops massacred unarmed Chinese personnel on the islands. Regardless of who ignited the crisis, once the firefight erupted between the opposing forces on land, both nations' naval assets were ordered to open fire in support.

The PLA-N capitalized on its initial victory and continued to carry out a systemic campaign to clear the Paracel Islands from Vietnamese troops with the participation of hundreds of Chinese troops and the purposeful application of airpower, against which the RVN had no adequate countermeasures. The remaining RVN garrisons were pacified and China secured control over the entirety of the Paracel Islands group.

In the conflict, we can see both a military build-up (affecting the balance of power projection) and attempts to seize land features (affecting the territorial balance). The conflict descended into an armed confrontation as both Vietnam and China responded militarily to reassert their weakening position in the dispute: Vietnam to the incursion of fishing vessels and China to Vietnamese troop landings. The conflict ended with a Chinese military victory, gaining control over the entire island group. This is an important reminder that in this context balance of power refers to the assets in the disputed territory, not a total balance between the two countries. While the local balance initially favoured South Vietnam, China could capitalize on its overall military power to execute a build-up that overpowered local Vietnamese forces.

But there are instances where much subtler issues can also represent a negative shift, resulting in escalation. This usually comes to the bafflement of the other claimant states. The Sino-Japanese competition over the Senkaku/Diaoyu Islands presents a couple of examples of this. In 2010, there was a collision between a Japanese Coast Guard (JCG) vessel and a Chinese fishing vessel in disputed waters around the Senkaku/Diaoyu Islands. Tokyo argued that the Chinese vessel intentionally rammed the JCG vessel and arrested its crew and captain. Beijing strongly protested the move and tensions escalated rapidly as China resorted to diplomatic sabre rattling and the threat of economic sanctions was rumoured. In 2012, Tokyo purchased three of the disputed islands from private owners. This was done in an attempt to de-escalate tensions, as the nationalist mayor of Tokyo sought to purchase and develop the islands, clearly meant as a provocation for China. Nevertheless, China responded with escalatory steps, including significantly stepping up its physical presence in disputed water. At the height of the conflict, there was an incident of Chinese vessels locking targeting radars on Japanese ones, which could have been easily interpreted as a precursor to an attack and could have led to the onset of hostilities.

At first sight, these escalations seem puzzling: No meaningful change in the dispute has taken place. After each incident, Japan did not enjoy any more sovereign control than it did before. The first one is largely a law enforcement matter, while the second one, although a change in ownership occurred, did not represent a change in sovereignty. Yet, China has resorted to escalation in both instances. Fravel argues that if a state's claim is weak enough, even administrative declarations are viewed as threatening.[7] The main issue for China in both instances has been the fact that Japan has exercised sovereignty, reinforcing the appearance of Japanese administrative control. China's claim to the Senkaku/Diaoyu Islands, both legally and as defined by Fravel, is very weak. It possesses no land in the disputed region and its ability to project military power into the region is very limited due to an unfavourable naval balance with Japan. As such, even minor shifts are viewed as significant, since China has very little ground left to lose. In both instances what Japan viewed

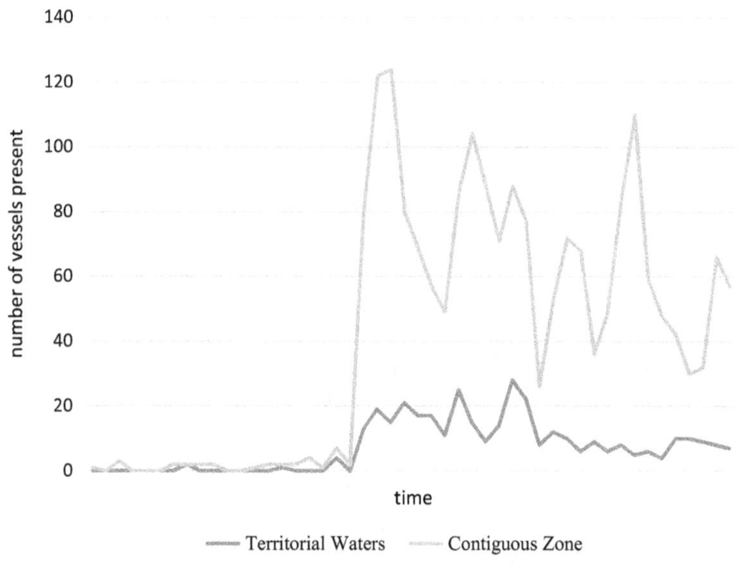

Figure 10.1 Escalation of Chinese physical presence in disputed waters (2011–2014)

Source: Ministry of Foreign Affairs of Japan (07 March 2016), "Trends in Chinese Government and Other Vessels in the Waters Surrounding the Senkaku Islands, and Japan's Response". Accessed on 11 April 2016

as routine day-to-day activity was perceived as a highly threatening negative shift by China that necessitated a strong response.

There are two ways in which military power can be used to address a negative shift in bargaining power: To seize territory or to signal resolve.[8] Seizing territory is straightforward. By gaining control of more of the disputed area, a state can directly increase its bargaining power. In contrast, signalling resolve is tied to deterrence. Deterrence can come into play in two key ways. On the one hand, by signalling resolve that a state is willing to protect its claims militarily, one can deter an opponent's military build-up by raising the perceived costs of a military resolution to the dispute. With sufficient deterrence in place the opponent is disincentivized from continuing the build-up and is incentivized to de-escalate by drawing down troops. The Philippines is attempting such a strategy against China when it is pursuing a military build-up of its own in the disputed area to deter China from *de facto* occupying contested reefs. Through this the negative shift in bargaining power is countered. But, on the other hand, deterrence can also be directed against other actions. By signalling that provocations will be met with force, a country can deter actions other than a military build-up that it nevertheless perceives as threatening. This is how China tends to use its military and other hard power assets on the East China Sea, where it faces a significantly negative balance of bargaining power. China pursues the aforementioned build-up in physical presence to signal Japan that exercises of administrative control are viewed as unacceptable and it will lead to escalation in the dispute. In the 2010 fishing captain incident, while the focus was on diplomatic sabre rattling, China successfully used escalation to secure the release of the captain, essentially leading to Japan forgoing the exercise of its own administrative control over the territory. In China's eyes this was sufficient to address the perceived negative shift in bargaining power and de-escalation followed.

But if military or hard power assets can be used as a tool of signalling, Wiegand questions whether it is possible to transfer reputation between disputes, meaning that escalation occurs in one dispute to signal an opponent in another dispute.[9] After all, a reputation for resolve should be translatable across multiple disputes.

In a system affected by imperfect information, reputation is highly valuable. Since states can never be sure of the others' intentions and key decision-making factors such as the balance of power being more of a perception than a tangible indicator, states must rely on the reputation of other states to make judgements. If a state has a reputation to be aggressive, it is reasonable to regard it with concern, even if it is conciliatory. If a state has a reputation to be a pushover, then one can demand more in negotiations or disputes. But if a state has a reputation of resolve to resist demands, then a more diplomatic approach is needed. Thus, how states act in a territorial dispute will not only affect that dispute itself, but also affect the reputation of the state in general. Appeasement and weakness will increase demands for further accommodation. Resolve and strength will deter such demands.

Wiegand argues that it should be possible for states to transfer reputation between disputes.[10] By choosing to escalate in a dispute, a state is signalling all other opponents even in other disputes that it won't accommodate demands. In reverse, failing to do so will signal claimants in other disputes of the potentially weakened state of the country. Thus, building up a reputation is essential. Based on this logic, one could argue that China partially pursues its claims assertively on the East China Sea to signal South China Sea claimant states. If Beijing is willing to be assertive against a major power such as Japan, it is reasonable to assume for ASEAN states that the same will apply to them, leading to seeking more conciliatory approaches.

Military force or other hard power assets in this scenario are used as costly signalling. Threats are useful, but for them to be credible sufficient resolve needs to be demonstrated to carry out the threatened consequences. Schelling argued for the need of such demonstrative action in his thesis on the diplomacy of violence.[11] Thus, a state might choose to pursue escalation and a limited military clash not to meaningfully alter the status quo directly, but to demonstrate its own resolve to follow through with its threats of violence. As such, violence becomes part of the complex psychological manoeuvring of deterrence and coercion.

Overall, the escalation of territorial disputes can contribute to the occurrence of war in straightforward ways. If there is a balance within the dispute, then states are incentivized to compromise to resolve the dispute as neither has a good chance to rely on military power to get the upper hand. In contrast, if there is an imbalance, then the stronger state can use military power to gain advantages or even resolve the dispute without having to seek a compromise with the other claimants. As such an imbalance is inherently threatening to any state that is suffering the negative shift. This incentivizes these claimants to counter these shifts, if necessary, through costly military signalling, again potentially leading to the outbreak of hostilities. Although they face an already negative balance, waiting further would only lead to the deterioration of the situation. In their strategic calculations facing an opponent today under somewhat unfavourable conditions is still better than facing them later under significantly worse conditions. Either way, territorial disputes represent a crucial focal point for competition that often leads to a military confrontation.

In the dynamics of the territorial disputes, one can observe a replication of the realist logic of balancing and imbalances leading to conflict. The system, whether a particular territorial dispute or the international system in general, is most stable when there is a balance. A relatively equal distribution of power forces states to negotiate and compromise since the promise of getting more through use of force is removed from the table by the opponent's ability to resist. But if there is an imbalance, states face an incentive to gain more through the use of force, pursuing their key interest. Thus, a keen awareness of the balance of power and the importance of balancing behaviour for the creation of a peaceful international system is highlighted once again.

Diversionary wars

Traditionally, wars are analysed as interstate conflicts. As such, their motivations are expected to be rooted in the international system. However, domestic factors can also motivate the decision to go to war. In these cases, it is less about mutual animosity, and more about the particular internal politics of a country. This idea is largely expressed through diversionary war theory, which presents the hypothesis that certain governments may choose to engage in escalatory policies and war to serve domestic political goals, primarily regime survival.

Psychologically, in times of crisis people tend to rally around the state. The existence of a clear external enemy puts internal dissent on hold. This is especially true in the case of a war. The survival of the state is potentially at stake and some of the citizens are asked to make the ultimate sacrifice in the service of the nation. Thus, engaging in internal debates is often viewed as unpatriotic and needlessly divisive. Whatever question may be about the government, they have to wait until the crisis has been overcome. It is not hard to see how an unscrupulous government could see potential in such a psychological response.

Every ruler, no matter how autocratic, is accountable to the people to a certain degree.[12] And any leader who has forgotten this has found its time holding the highest office being cut short, sometimes violently. The government needs to maintain legitimacy, at the very least with its key selectorate. The key problem is that it is very difficult to maintain legitimacy, especially over expanded periods of time. Even more troublingly, seemingly good performance alone is not a guarantee for maintaining legitimacy and political power. Only 45% of Organization for Economic Co-Operation and Development (OECD) country politicians get re-elected, despite supposedly being some of the richest and best governed nations on the planet.[13] This creates incentives to limit democratic pressures on the government, which in turn also undermines the government's legitimacy. If the government loses its legitimacy, it is likely to face protests, strikes, or even an armed uprising. Thus, maintaining legitimacy is a key survival interest for any regime. And since the violent collapse of regimes usually does not bode well for their leaders, it is often a literal life or death question.

Diversionary tactics present a good opportunity to deflect criticism or distract attention from the government's shortcomings, as well as take advantage of "rally around the flag" responses to a threat. Designating an internal or external enemy is important. They can be blamed for anything that goes wrong and offer a clear target for people to concentrate their anger or frustration on. Sabre rattling and aggressive rhetoric are often sufficient to flame the unifying forces of nationalism. However, diversionary wars offer one more thing to leaders: The opportunity to demonstrate competence. Being a wartime leader is regarded highly as it greatly increases the prestige and reputation of a leader. To stand at the nation's darkest hour and to guide it through is considered to be an achievement that pays dividends long after the actual conflict. Some of the best remembered statesmen have been war-time leaders, whether one is talking about Churchill, Roosevelt, or De Gaulle. The roots of the Chinese Communist Party's (CCP) legitimacy are in its revolutionary movement and its role during China's resistance against Japan, with many leader's personal political fortunes closely tied to the role they played during that time. Showing leadership during the crisis can grant a lot of leeway to a leader in future political decision-making and make up for criticism and domestic political challenges.

China presents a good example of how diversionary tactics can be intertwined with internal political legitimacy. One of the key challenges facing the PRC is the weakening legitimacy of the CCP. During the Cold War, there were four main sources of legitimacy for the CCP: (1) The Cold War ideological context and threat of superpower wars, (2) the Party's Marxist–Leninist class warfare ideology, (3) the ongoing conflict with the Kuomintang in Taiwan, and (4) Mao's cult of personality. However, over time these foundations have weakened. By 1989, Mao's cult of personality was gone and his revolutionary Marxist faction was defeated and disowned by the CCP, the Cold War and the threat of invasion was winding down, and Deng's economic liberalization has deeply damaged the party's Marxist–Leninist credentials due to growing inequality. Thus, weakening legitimacy culminated in the 1989 Tiananmen Square protests that sought an end to the Party's monopoly on power and the establishment of a competitive political system. The Party has responded with repression, but it was clear from the examples of the former Eastern Bloc that such a system would not last. It would also destroy all the fruits of Deng's economic reforms. As such, the party has sought alternate sources of legitimacy: (1) Economic development to deliver increasing living standards and (2) a patriotic narrative (discussed above) to replace much of its Marxist–Leninist ideological legitimacy. The system worked very well. This two-pronged system, combined with growing internal surveillance, has halted the challenge to Communist rule in China and the PRC is one of the few countries where the communist regime survived the fall of the Soviet Union.

However, these only provided a temporarily respite for the Party. China's economic development is slowing down, decreasing the Party's ability to deliver economic development. China has been reaching the limits of its economic model and experienced previous shocks as export-oriented growth was stifled and the construction sector struggled. Essentially the PRC is entering a period similar to that which Japan experienced in the 1980s, when it too reached the logical limits of its own economic model. This pushes the Party to be more reliant on its patriotic credentials. However, those too are suffering weaknesses. As a great power the idea that China is under constant threat holds less credibility and much of the historical narrative has weakening relevance as distance increases between the present and China's war of resistance against Japanese aggression. Based on this, the current trends in Chinese politics are hardly a surprise. One can observe the re-emergence of some elements of a personality cult with Xi Jinping, tying parts of the state's legitimacy to his own person. At the same time, Xi's China has been stepping up antagonism with other regional powers, especially Japan, which is most relevant to this section. China engages in diplomatic sabre rattling and escalatory policies against Japan partially motivated by domestic reasons. By providing an external antagonist, China offers an outlet for domestic frustrations to be channelled away from the Party. At the same time, the presence of an external antagonist reinforces the Party's legitimacy as the vanguard of China's rise and the need for internal unity.

The key drawback of these policies is that they limit the Party's policy options in the long run. Since the CCP is investing a lot of political capital into its standoff with Japan, backing down or adopting a conciliatory tone is not an option without losing face and credibility back home, potentially triggering a new legitimacy crisis. This means that escalation can only be restrained by Japan, as China is forced to match any moves Japan makes. At the same time, these diversionary tactics have their own escalatory tendencies. As China pushes parts of its population to adopt more nationalist/patriotic tendencies, sabre rattling won't be enough in the long term. The population will demand action, especially as the power of China continues to grow. Inaction will be viewed as weakness and trigger criticism. Many of the incidents concerning the territorial disputes have been caused by nationalist activists from either side taking matters into their own hands. As such, trying to harness these diversionary tactics is a double-edged sword. They can distract the population for a while, but they also make it increasingly harder to avoid a confrontation.

Depending on the severity of the legitimacy crisis, a state might decide to actually go to war as a result of these diversionary efforts. Sabre rattling is useful, but they have limited effect on silencing dissent. Their effect is not comparable to the onset of actual war, where even the ardent critics of the government would need to prioritize conflict resolution.

One example of a war where diversionary factors were a component is the 1982 Falkland Islands War. The military junta in Argentina has been facing mounting internal dissent due to repressive policies and the general ineffectiveness of the government. In response they sought to capitalize on the long-standing territorial dispute with the United Kingdom (UK) as a diversionary tactic and an opportunity to demonstrate competence. The war was expected to be fairly simple. The UK was far away with minimal military presence in the region. The UK was also suffering from economic issues and as the islands offered minimal strategic value to them, Argentina expected the UK not to respond strongly. Achieving a quick victory and reclaiming historical territory would have boosted the legitimacy of the government. Unfortunately for the junta, PM Thatcher did respond militarily and managed to win the conflict, boosting her own legitimacy in the process, while the junta's legitimacy suffered a further blow.

One of the key issues of diversionary considerations affects one's ability to resist a war. If the roots of the conflict are in domestic politics, rather than international dealings, it becomes so much harder to deal with, especially as one has limited ability to influence the internal processes of a country. This is a situation ASEAN countries and Japan have to grapple with in the context of China. Japan tried its extensive strategy of engagement. Similarly, ASEAN tried a strategy of enmeshment to mitigate Chinese assertiveness.[14] Both strategies have largely failed. Since China's assertive foreign policy, especially against Japan, is often tied to its own internal dynamics and the weakening legitimacy of the Chinese Communist Party, foreign policy efforts have exerted limited influence. In the case of Japan, the situation has gotten to the point that, despite its anti-militarist identity, Tokyo had to adopt a policy of classical direct military deterrence. It is only by manipulating the balance of power that Japan can effectively respond to China's actions on the East China Sea. This however creates a problem of its own. Increasing Japan's military might as a response reinforces justifications in China for designating Japan as an opponent, as well as generally intensifying the security dilemma between the two countries.

The ideas of diversionary wars are important because they demonstrate that one often cannot make a clear-cut separation between domestic and foreign policy. Domestic and foreign policy, as well as policy and organizational interests, are inherently intertwined. Sometimes decisions are made for political survival rather than national interest. The considerations muddle our clean understanding of rational state behaviour, in which supposedly national interests rule supreme. But it is important to remember that rationality is a normative concept. It is how things should be. And it is useful for modelling interstate relations. But it is never 100% accurate. North Korea's ongoing military posturing and aggressive acts against South Korea are just as much about the country's internal politics as about demonstrating resolve as part of a deterrence strategy against external opponents. Some of the most puzzling international conflicts come down to internal political calculations, and as such they cannot be ignored when analysing a state's capacity to go war.

Ultimately, one should never underestimate a leader or regime's willingness to hold onto power. And where there is a will, there is a way, even if it requires the commitment of the country to an armed confrontation.

Notes

1 Rasler, K. A., and Thompson, W. R. (2006) Contested Territory, Strategic Rivalries, and Conflict Escalation. *International Studies Quarterly* 50(1), pp. 145–167.
2 Fravel, M. T. (2007) Power Shifts and Escalation: Explaining China's Use of Force in Territorial Disputes. *International Security* 32(3), pp. 44–83.
3 Rasler, K. A., and Thompson, W. R. (2006) Contested Territory, Strategic Rivalries, and Conflict Escalation. *International Studies Quarterly* 50(1), pp. 145–167.
4 Ibid., pp. 146.
5 Fravel, M. T. (2007) Power Shifts and Escalation: Explaining China's Use of Force in Territorial Disputes. *International Security* 32(3), pp. 44–83.

6 Ibid.
7 Ibid.
8 Ibid.
9 Wiegand, K. E. (2011) Militarized Territorial Disputes: States' Attempts to Transfer Reputation for Resolve. *Journal of Peace Research* 48(1), pp. 101–113.
10 Ibid.
11 Schelling, T. (2008) *Arms and Influence.* Yale University Press: New Haven.
12 De Mesquita, B. B., and Smith, A. (2011) *The Dictator's Handbook.* Public Affairs: New York.
13 Ibid.
14 Goh, E. (2007) Great Powers and Hierarchical Order in Southeast Asia: Analyzing Regional Security Strategies. *International Security* 32(3), pp. 113–157.

Part III

Conclusion

The reason we fight

Understanding why a country might go to war is important for two reasons. First, it can provide guidance on whether a policy issue at hand is within the realm of war or not. Second, by understanding the mechanics of war, one can use this to anticipate the moves of other countries. This is crucial for effective preventative policy formulation since it is always done in a game theory environment where there is a series of reactions to any steps.

As Part III shows, the reasons for war are complex and often multifaceted. An imbalance of power, at least a perceived one, provides the most fundamental condition that permits a war to occur. Without such an imbalance states are in equilibrium and there is limited chance of victory, often rendering conflict to be irrational to attempt. This is based on fundamental realist principles. States are rational entities basing their decisions on cost–benefit–risk analysis. If there is no sufficient chance of victory, even offensive structural realism would caution against the use of force. Thus, a perceived advantage in the balance of power, or a fear that a conflict is unavoidable and would be fought under worse term later, is a precondition for war. But this alone does not capture the full picture.

While balance of power permits the occurrence of a war, it hardly explains it. Most states are not guided by the pure balance of power logic described in defensive or offensive structural realism. This necessitates the introduction of additional factors. These are factors that actually push a state to make the decision to go to war. Factors such as geopolitics, culture, identity, or hegemonic rule-setting fundamentally affect what state interests are in the international system, how important they are, and whether a state is willing to use force to get it. To understand a state's capacity for war, one needs to understand the cognitive framework in which that decision is made – at least in a normative sense. To paraphrase Sun Tzu, it is essentially to understand the opponent in order to develop effective policy measures, whether one is considering the offensive application of force or seeking to develop an effective strategy to deter it.[1] The theoretical context of war is immensely complex and as such one's model has to incorporate a number of factors to produce reliable results. With this we understand why a state might chose to go to war. Part IV focuses on reaching the rational decision to pick up arms against another state in the international system.

Issues for further thinking

This section offers a few questions and scenarios that one can explore to further reflect on the key points made in this chapter. For each question, this section provides some key details and points to consider but it ultimately does not seek to provide a definitive answer.

Is underdevelopment a security threat?

Barnett has argued that disconnectedness is a security threat to the developed world. That despots and rogue regimes use disconnectedness to maintain their power by keeping their citizens poor and uninformed, artificially keeping out the benefits of globalization. By "connecting" these countries to the global order one not only increases global security but also allows these citizens to live their lives to their full potential. On the surface this is a compelling argument for a type of benign hegemony that was expressed in the Bush administration's political ambitions.

However, this perspective faces off against a number of critical arguments. Huntington, who has been discussed in Chapter 9, has argued that the clash of civilization is partially a result of local resistance to the creation of a global culture based on Western norms. Huntington cautioned against equating modernity and development with Western civilizational values in interpreting globalization. Another critical angle is more economic focused. Wallerstein has argued about the existence of a capitalist world economy, an all-encompassing global economic system that is based on the capitalist mode of production and division of labour. Within this division of labour there are advanced core countries and poor peripheral countries. Poverty and lower levels of development are not a result of accidents or historical misfortunes but are a feature of the system of the world economy. Capitalizing on the earlier developmental debates Wallerstein argues that these countries are not undeveloped (simply not having realized development) but are underdeveloped, meaning that they are systematically prevented from realizing their development by the exploitative nature of the capitalist world economy.[2]

This raises significant questions about the securitization of underdevelopment. Theories such as Barnett's core-gap thesis view inequality and developmental issues as largely a result of rogue elements and their political machinations, rather than confronting potential systemic issues at play. This was evident in the Bush Administration's failure in Iraq. The Administration assumed that the primary obstacle to democratization was a rogue regime seeking to maintain power and that once removed the introduction of democracy would be a straightforward process. In retrospect, arguably the Saddam regime was more a symptom of deeper issues rather than the disease itself, however horrible it might have been. The Bush Administration has failed to appreciate Collier's point on how democracies are more than the simple act of voting: They are robust institutions that can sustain majority rule.[3] In the absence of such institutions, the introduction of democracy in such an institutional vacuum is doomed to act as a destabilizing force, as it did in the case of Iraq where the Western-supported government struggled to maintain legitimacy and control over the country.

From this book's perspective it means that there are significant questions concerning why one goes to war. The road to hell is paved by good intentions. Selecting improper reasons to go to war, however well-meaning, is likely to lead to terrible results. At the same time, the issue also highlights the threat of improper analysis. Wars are fought to achieve a post-war political outcome, mostly using war to remove obstacles to that political reality. But what happens when the obstacles are identified incorrectly? By securitizing underdevelopment, and attributing the outcome to rogue regimes, rather than adopting a focus on systemic factors, one is likely to attempt to wage a war for an objective it can never accomplish. This would render all the sacrifices and tragedies of war meaningless. Before one decides to go to war, it is crucially important to be clear on whether the right targets for action have been identified.

Can the political, cultural, and physical maps be reconciled at this point?

After several thousand years of human history and some 400 years since the Treaty of Westphalia (1648), a large part of which was spent waging war and redrawing national boundaries, there is a significant disconnect between maps. Nations that make up the political map are largely arbitrary constructs, subject to nation building or expansion and contraction based on the nation's fortune at war. The current borders of nations hardly correspond to either the civilizational distribution of people or the ideal geopolitical boundaries. At the same times, nations have invested significant ideational capital on portraying their borders as something sacred and inviolable. This raises the question of whether the differences between the political, cultural, and physical maps can be reconciled at this point or whether humanity is forced to live with the legacy of a long history of warfare and redrawing of boundaries. One has to ponder whether it is even possible to create peace as long as such disconnects exist.

Nation building is a complicated task. As Benedict Anderson argued, the nation is primarily an imagined community as the scope of it is beyond the practical understanding of any one individual.[4] This sense of an imagined community is achieved through the creation of national symbols. These can be flags, the national anthem, key landscapes (e.g., Mount Fuji), and so on. One important element is the boundaries of the country. Maps are frequently displayed and from an early age most people could visualize the physical shape of their nation. That image is created as the "whole" and any deviation from it would be a "wound" on the nation. A main exception to this is the group of countries whose boundary image is already presented as wounded. Hungary, for example, often presents its current borders superimposed over its pre-Trianon borders. Most people can not only visualize the contemporary borders but the pre-World War I ones, creating a sense of loss in national image. The current borders are still viewed as inviolable, but the superimposition creates a sense of potential rightful recovery. To a certain extent Chinese nationalist politics also "plays" with the image of borders. They can highlight the loss of territory to Western imperialism and their recovery or display aspirations for the recovery of "lost territories", especially in the context of maritime territorial disputes or Taiwan. The larger borders are viewed as the legitimate, inviolable one, expressing a national wound that needs to be rectified.

It is not hard to see how such political imagery can be a tool to alter the political map to bring it into more harmony with the physical or cultural one. One example is Hungary's "red map" issued at the end of World War I where it sought to alter the border arrangements in Trianon by highlighting the ethnic distribution of the country, seeking (unsuccessfully) to attach Hungarian majority areas that were demarked to be transferred to neighbouring countries. How a country could use these to correct the political map through expansion is easy to see. However, Kaplan makes the argument that states expand and contract to accommodate the physical map.[5] Sometimes it is in a country's interest to give up territory to become more secure by conforming to national boundaries and cutting lose territory that would be a source of insecurity. Similarly, it might be in a country's best interest to grant independence to groups within the country that have been a source of domestic insecurity. However, contraction is much harder to justify, especially in the context of nation building and imagined communities. There are examples when countries are willing to give up land: China has settled a number of land border disputes through offering significant concessions to other countries in order to conform to physical geography. However, major adjustments are relatively rare, and countries are rarely eager to give up significant territory or people, even if it would make them more secure.

Nation building is often a process of taking an artificially delimitated territory and people and attempting to forge a cohesive unit out of it. If effective, borders become viewed as inviolable, which is good for mobilization and maintaining the status quo. But sometimes it stands in the way of a state recognizing other strategic realities. It can push states towards aggression to recover lost territories. And it can prevent states from recognizing that sometimes it is best to let some territories go.

One potential scenario is the Falkland Islands. The islands are part of Great Britain and culturally consider themselves British. However, they are off the coast of Argentina and are claimed by the country. The islands are remnant of Britain's expansive colonial empire. In 1982 Britain had to launch a complex military operation, sending a fleet halfway across the planet, to recover them when Argentina invaded them. This exposes the conflict between national and geopolitical considerations. Politically and culturally it makes sense that the islands would be part of Britain and that it would protect it through military means if necessary. However, from a geopolitical perspective it makes limited sense that such remote possession would be held, especially as this makes their defence extremely difficult. From a utilitarian perspective it makes sense that one would think that it might be best for Britain to give up such possessions. That was certainly what the Argentinean junta was counting on. However, as the outbreak of a war showed, they underestimated just how much this is an unsatisfying answer from a political and nationalist perspective.

Are concessions a good strategy to break escalatory cycles?

If one looks back above to Fravel's theory concerning the escalation of territorial disputes, it shows that escalation can be caused by a negative shift in bargaining power. A military build-up or even some administrative declarations can make an opponent perceive its position as weakening and thus trigger an explanatory response. If the opponent's position is weak to begin with, this trend can be amplified, and even minor events can trigger a hostile response. Cycles of escalation are dangerous periods that can lead to a war neither party wanted. This raises the question of how to break such cycles. One approach could be to offer concessions in such a situation in order to reinforce the position of the opponent and stop the perception of the negative shift to de-escalate. In many instances the escalatory party does not seek a resolution to the dispute, but simply an acknowledgment to the continued validity of its claims.

There is a school of thought that appeasement is a sign of weakness and that it will always invite more demands as it emboldens aggressive states and confirms the effectiveness of their coercive behaviour. However, a no concession strategy is questionable while maintaining peaceful coexistence. International relations are a constant give-and-take between states. Bargains are struck and concessions are offered for future benefit every day. Coercion itself is a process of bargaining, if a violent one. Sometimes it makes rational sense to make concessions rather than to risk war, especially if the demand is of far lesser value than the cost of a potential conflict. While there is a value to building a reputation of not giving in to coercion, sometimes it offers practical utility to simply defuse a situation through concessions.

At the same time, even if a state is sometimes willing to engage in concessions, there is no guarantee that it will work. A good example of this is the decade-long dispute between China and Japan over the Senkaku/Diaoyu Islands. Japan has traditionally tried to engage in a foreign policy that tries not to antagonize China, preventing a negative shift in perception described by Fravel. Japan maintains a self-imposed moratorium on developing the islands and in 2012 when the nationalist mayor of Tokyo tried to purchase the islands the government stepped in to defuse the situation to the best of its ability under Japanese law. It nevertheless triggered the worst escalation in tensions in years. While Japan viewed the government's purchase of the islands as a defensive move to de-escalate tensions and preserve the status quo, China viewed it as a strong display of effective control, signalling the weakness of China's claims. This is underpinned by the fact that Japan denies the existence of a dispute, trying to walk the tightrope of not antagonizing China while advocating for its own territorial interests.

Ultimately, concessions are a difficult question. While a country often has an interest in preventing escalation, it also has an equal or stronger interest in preserving the strength of its claims. Concessions often can only be made by strengthening the claim of the opponent over the territory, even just by simply acknowledging that they might have a legitimate claim on the territory, which runs contrary to one's interest of ultimately possessing the territory. For example, when in 2010 Japan released the fishing captain who rammed the Coast Guard vessel due to pressure from China, it was a weakening of Japan's administrative control. Similarly, when Japan chooses not to confront vessels around the disputed islands, it signals a weakening of its effective control, because it cannot freely act as the sovereign of the islands. And this is the intended outcome of China's strategy. At the same time, confronting these vessels or refusing to repatriate the vessel would have caused a significant spike in tensions and potentially significant economic damage due to potential punitive measures form China.

Overall, managing escalation is a delicate art. Whether to offer concessions or hold firm should be decided on a case-by-case basis and the decision is often affected by factors beyond the immediate dispute, such as reputational consideration.

Notes

1 Sawyer, R. D. (2007) *The Seven Military Classics of China*. Basic Books: New York.
2 Wallerstein, I. (2006) *World System Analysis: An Introduction*. Duke University Press: Durham.
3 Collier, P. (2009) *Wars, Guns and Votes: Democracy in Dangerous Places.* HarperCollins Ebooks: New York.
4 Anderson, B. (2006) *Imagined Communities: Reflection on the Origin and Spread of Nationalism*. Verso: London.
5 Kaplan, R. D. (2013) *The Revenge of Geography*. Random House: New York.

Part IV

The rational decision to go to war

The previous two parts of this book have engaged with the occurrence of war in the international system from a largely theoretical perspective. They discussed key theoretical approaches and provided critical engagement to contextualize these theories and understand their strengths and weaknesses. Part IV moves in a somewhat different direction. The main objective of this part is to create a conceptual framework for making the decision to go to war. The analysis below primarily relies on the construction of a rational choice decision-making model and applying it to the question of war. The adoption of a rational decision-making model is in accordance with the key theoretical assumptions of this book: From a normative perspective we assume the system to be rational. This part visits the different stages of the rational decision-making model and explores how one can make key determinations about war. The abstract discussion on the construction of the model is supported by real-world examples for context. Ultimately, war is always a tragedy. But this should not be taken as an excuse to forgo critical engagement with the decision to go to war. Understanding the nuance of deciding to use force can provide critical insights, even if one only seeks to prevent the occurrence of war in the international system. It is the very magnitude of the tragedy of war that necessitates a thorough understanding of the subject.

The discussion on the rational choice model is divided into a few key topics:

1. The concept of rationality and bounded rationality as a foundation.
2. Feasibility testing: Determining whether war is a feasible policy option.
3. Cost–benefit–risk analysis: Establishing the "attractiveness" of war as a foreign policy option.
4. Comparison and selection: Comparing the relative attractiveness of war to other policy avenues and making a rational choice on the policy avenue based on this comparison.

Figure PIV.2 illustrates the flow of the rational choice model. This contains all the key components needed to construct a working rational choice model for policy analysis. Part IV concludes with going beyond a singular policy choice to contextualize policy decision-making from a more game theory-oriented perspective and to explore the potential quantification of these calculations.

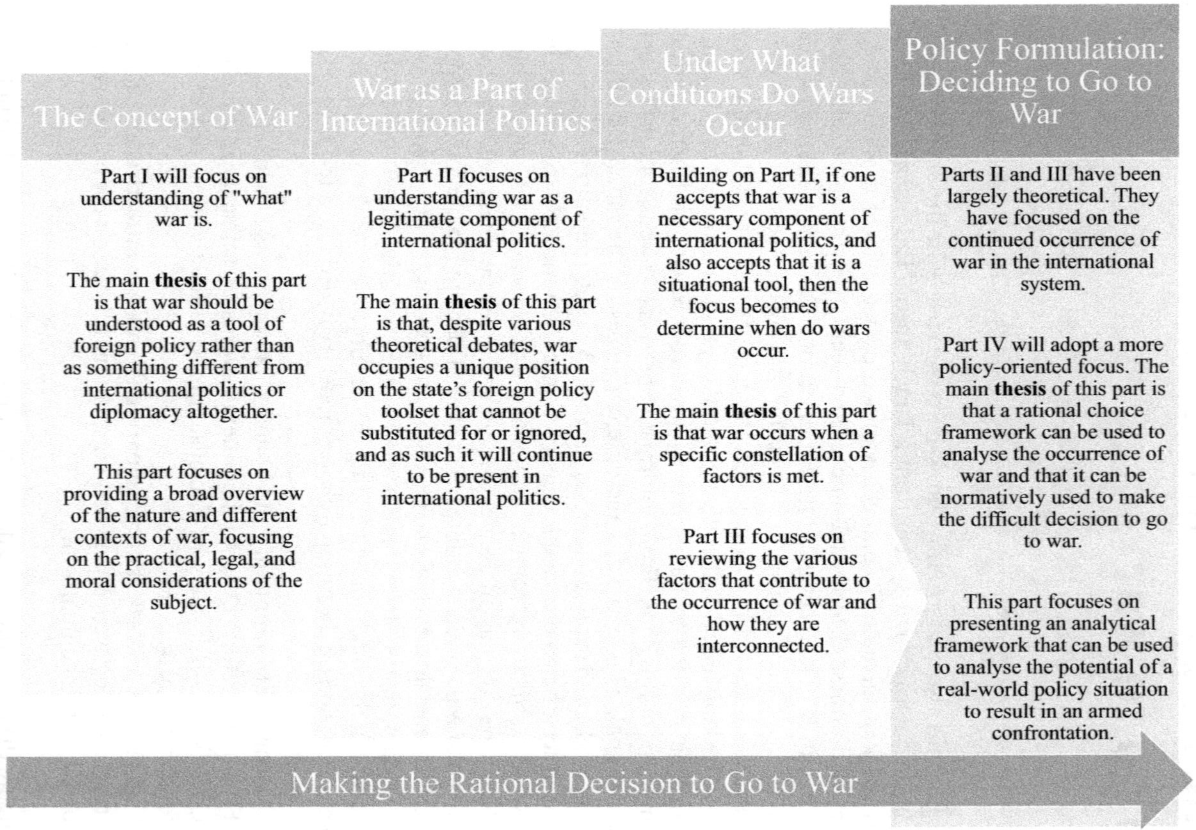

Figure PIV.1 Part IV's place in the overall arch

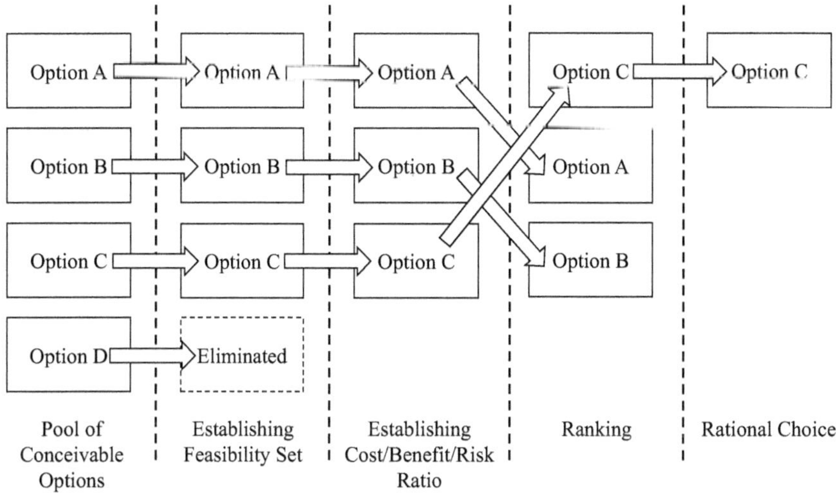

Figure PIV.2 Rational Choice Model

11 Rational choice modelling for war

This chapter focuses on the construction of the rational choice model this book proposes to use to critically analyse the occurrence of war, especially in a normative or anticipatory context. As the term "modelling" implies, the goal is to create a representation of the decision-making process into which information can be fed to reach conclusions. By necessity this will be a simplification of the infinitely complex reality one inhabits. But if constructed with sufficient care it can nevertheless maintain practical utility for policy analysis and formulation. The chapter is divided into two main parts. The first part examines the concept of rationality for International Relations, emphasizing the need for a bounded rationality-based approach. The second part focuses on the framework of the rational choice decision-making model, walking through the steps of the process. The key concepts highlighted in this chapter, such as feasibility or cost–benefit, are developed further in the following chapters.

Rationality in international relations

Looking at the world in its current stage, it is sometimes hard to agree with the assertion that the international system is inhabited by fundamentally rational states. The very topic of this book is the occurrence of organized, and potentially large-scale, violence within the system. It is relatively easy to look at the actions of major world leaders and question of whether they are of sound mind when making seemingly random or counter-intuitive decisions. Countries are willing to risk to or actually go to war over seemingly worthless territory or in instances when negotiations could resolve issues. Global cooperation breaks down in a time of need. Countries pick petty fights over seemingly small issues instead of focusing on larger ones. The planet itself is heading towards its limits of sustaining human habitation, yet collective action continues to be missing or on some occasions positive developments are actively reversed. It is understandable that some would throw up their hands in exasperation and question the rationality of the human enterprise.

However, many of the grand International Relations theories argue that the international system is inhabited by fundamentally rational actors. This is treated as a central assumption that underpins many of their theoretical tenets of the school of thought. Mearsheimer specifically argues that one of the five key assumptions of offensive structural realism is that states are rational actors.[1] The idea of rational state behaviour runs deeply through the realist tradition of International Relations. But rationality is also central to many of the liberal traditions, for example, economic interdependence theory. There are two key ideas to note when it comes to rationality in the International Relations theory: First, rationality is necessary from an analytical perspective but it is somewhat of a simplification of the model. Second, while most International Relations theories are built on strong explanatory foundations, they are fundamentally normative in their theoretical scope.

The idea of constructing models has been brought up before in this book. Reality is infinitely complex; thus, much of International Relations as an academic discipline is taking place within the confines of constructed models that emphasize and de-emphasize different parts to reveal underlying mechanics or anticipate behaviour in order to offer foreign policy advice. The assumption that states will behave in a rational way is such a modelling issue. States *mostly* behave in a rational manner. But the irrational inherently resists analysis. From an analytical perspective, especially from a predictive angle, there is not much one can do with the irrational, other than recognize that they are irrational. In a game of rock-paper-scissors, one can develop a strategy as long as the other player acts within the confines of the game. If the opponent plays banana (a completely inappropriate response outside the confines of the game) or plays rock 700 times in a row (an inappropriate response within the confines of the game), then no amount of planning will work out. One could conduct all foreign policy under the assumption that the slightest conflict would trigger thermonuclear war, but that would simply paralyze analysis and decision-making. Thus, as an analytical shorthand, for modelling purposes the states are considered to be rational because what is rational can be understood, analysed, and predicted, and what is not, cannot.

The thought that grand theories such as structural realism or economic interdependence are fundamentally normative in nature might seem as a heretical notion. They built up tomes of explanatory research that shows how states conform to the behavioural patterns they prescribe. While they build on observations of behaviour, they prescribe how states *should* behave to be successful in the system. They do not diminish the agency of states to deviate from these principles. At its core, offensive or defensive structural realism, economic interdependence, geo-economic statecraft, smart power, or democratic peace all seek to describe ideal behaviour, how states can be successful if they understand the rules of the game. This is recognized by Waltz, for example, when he argues that systemic forces limit the number of potential successful strategies.[2] Importantly, there is a recognition that states can deviate from these paths, but they do so at their own peril. If anything, there is a tendency to dismiss rule-breakers as irrational, rather than to engage with a potential underlying rationality that led to their actions.

Overall, rationality is important because it creates patterns of behaviour: Given a stimulus one can anticipate how a state will behave. This is crucial, especially in forward-looking areas such as foreign policy. After all, today foreign policy is formulated to deal with the problems of tomorrow. A foreign policy seeking to clean up problems that already manifested will not be a successful one. Deterrence is a good example. When formulating a policy of deterrence, one seeks to prevent an opponent from adopting a certain foreign policy option – war – in the future. If one does not understand how they react to certain actions, then deterrence is impossible. Thus, while rationality represents a simplification of the model, and reduces its overall accuracy, it is a necessary one. It is a compromise for the sake of utility: It is not perfect, but it is the best one got.

The idea of rationality is also muddled by other issues. On the one hand, rationality in an international relations context is often confused with the common language usage of rationality. On the other hand, observes often fail to note on which level of analysis is rationality present. What appears to be irrational as a policy outcome on the international level might be the results of rational decisions from the decision-maker's perspective.

Rationality is often confused with making the "right" decision as defined by a moral or other framework. In this context, rationality simply means that the decision-maker weighs the costs and benefits of various options and selects the one that best achieves the goal. These calculations are not inherently affected by preconceived notions of preferences for any given option. If anything, rational choice is based on the idea of giving a fair chance to any potential feasible option. For example, Angell considered war to be an irrational activity due to its destructive nature and the limits of conquest.[3] This lines up with the general view that war is a moral evil. But when examining a rational decision-making situation, one cannot simply dismiss war without fair examination. Ultimately an agent can remain rational even if one disagrees with their decision, as long as the decision-maker remains within the analytical framework of rational choice. In these instances, it is likely that the observer lacks an in-depth understanding of the decision-maker and the decision-making situation.

Similarly, it is important to note that on the international level the state is considered to be a unitary actor, and this is another simplification for modelling purposes. Within a state, a lot of different influences coalesce into a foreign policy direction. Individual or organizational interests might push policy directions that are seemingly irrational on the international level. If a state chooses a war, with all its associated costs, instead of pursuing negotiations, one can draw the conclusion that rational decision-making principles have been violated since the state did not pursue the optimal course of action. However, it is likely that some component of the decision-making process has been overlooked. The previous chapter discussed diversionary considerations for war. It is possible that the decision-maker had additional interests attached to certain policy outcome, beyond the state's own interests. Ideally, this would not happen. But realistically speaking it is not uncommon. As discussed before, in the People's Republic of China there is a fusion of domestic and foreign policy, both being subordinated to the Chinese Communist Party's survival. This doesn't make the process irrational necessarily, in the sense that the behaviour of the country still can be analysed and anticipated to a sufficient degree. But it introduces additional components into the analysis that need to be accounted for. Such considerations are common in autocratic societies where there is less separation between the government and the state. But it is not unheard of in democratic societies either. It is often hard to make rational sense of the Trump Administration's policies without understanding who is in charge. Ultimately the state is an abstract construct and it has no will of its own. Its actions are driven by the minds of its constituent parts with varying degrees of influence on the state's overall actions. The unitary state is an analytical shorthand[4] that is useful for analysing the policy directions of a state. But when it comes to the actual decision-making process, one has to recognize the limits of such shorthands. One has to ponder that if decision-makers have such influence, while much of International Relations scholarship maintains the unitary model. Unfortunately for analytical attempts, the minds of decision-makers are inherently unknowable. One can make educated guesses based on certain signs that a decision-maker has a bias or interest for certain options, but in the absence of mindreading, these can remain highly subjective or speculative. Instead of trying to discern the minds of people, some of International Relations scholarship simply black-boxes these aspects of the decision making process, accepting the resulting reduction in predictive validity. This harkens back to the discussion on rationality in the beginning of this section: It is not perfect, but it is better than nothing for analytical purposes.

This leads one to the concept of *bounded rationality*. Bounded rationality was introduced into rational choice analysis to account for the shortcomings of a deterministic universal rationality. The central argument of bounded rationality is that, to understand the decision-making of an agent, one needs to be deeply familiar with the preferences of that agent. One can illustrate this with a simple example of consumer electronics. One can take a coffee machine that makes simple coffee for 25 dollars. If a competitor introduces a different machine that makes both simple coffee and fancy lattes for 30 dollars, a universal utility maximizing logic would argue that the second machine is the rational choice since for 5 extra dollars one can produce twice as many types of coffee. However, a bounded rationality argument would say that it is not that simple. Because if the decision-maker doesn't consume lattes, then there is no reason to buy the second machine and pay 5 dollars extra for a function one will not use. What the correct course of action is ultimately is fundamentally bounded to the cognitive framework of the decision-maker. Ignoring bounded rationality essentially means imposing a certain logic over the decision-maker. While it can work in a purely normative context offering recommendations, in an explanatory or predictive context it falters. Constructing a bounded rationality model requires a deep knowledge of the decision-maker, in this particular instance, the state one is looking at. It goes beyond the universalistic logic of International Relations grand theories, and requires the incorporation of factors such as culture, strategic doctrine, and/or internal politics into the decision-making model. It essentially requires a specialization in that particular country to the point that a reasonable understanding of their unique cognitive framework is gained. How bounded rationality can come into play emerges in the following sections, when the chapter examines the specific components of the rational decision-making process.

Overall, when one is talking about rationality, it means that there is an analysable and ultimately predictable structure to the state's behaviour. Predicting powers are always constrained by the fact that analysis is done on a modelled version of reality that utilizes strategic simplification and black-boxing. But that is a limitation one has to contend with. The next section will focus on constructing the basic framework of a rational choice model that will provide the structure for analysing the decision to go to war.

Rational choice model

Constructing a rational choice model provides a solid decision-making framework for both explanatory and normative purposes. The model has its origins in economics where it was used to predict consumer behaviour. However, if one accepts that states are mostly rational, then the model can be applied to analyse the behaviour of states in the international system. Using a rational choice model can be especially useful for anticipatory or normative research as it provides a solid framework to avoid falling into conjecture. As a formal model, rational choice has set stages to progress through before a decision has been made. It is important to note that rational choice is designed for singular policy decisions. To analyse a chain of policy decisions (or a multilevel game) one needs to expand the analysis into game theory. To construct a single decision rational choice, one needs to go through the following stages:

1. Understanding the decision-maker and its interests.
2. Identifying the potential options available to the decision-maker.
3. Narrowing down the field to feasible options.
4. Analysing the costs, benefits, and risks of each feasible option.
5. Establishing a ranking of options based on the relative cost–benefit–risk ratio of each option.

Once the analysis progresses through these stages, a policy option will emerge as a clear "winner". Adopting that option will be the rational choice for the decision-maker.

Phase 1: Understanding the situation

To paraphrase Sun Tzu, having an in-depth understanding of all aspects of a war is essential to victory. Naturally, before a decision can be made, one has to be clear on what one seeks to accomplish through that decision. This connects up with the long-running theoretical debate about what are the key interests of states in the international system. Realism provides a clear and straightforward answer: Survival. All states seek to perpetuate their existence within the international system. In an anarchic international system this is achieved through acquiring sufficient power to ensure their own security. Essentially the main objective of their policy decisions is to either increase their relative power or prevent a negative shift in the balance of power at the end of the day, whether one subscribes to offensive or defensive structural realism. In contrast, liberal theories emphasize the creation of a harmonious international system that allows for mutually beneficial wealth acquisition as the end goal of foreign policy. This gives an overarching frame to the rational decision-making framework.

However, understanding has to go deeper than this for two key reasons. The first one is the fact that "security seeking" is a fairly abstract goal. This overarching goal has to be translated into smaller pragmatic policy objectives. It is clear that a state seeks more security, but one has to ask how this can be accomplished in the international system. Most of the policy decisions a state needs to make in the international system are not clear cut. Issues such as whether one should pursue disputed territory, joint development, or alliances with certain states are common, yet whether they would have a positive impact on the state's security is not immediately obvious. One has to gain an in-depth understanding of the actual decision that needs to be made to determine its potential to contribute to the overarching policy goal.

China and the Senkaku/Diaoyu Islands presents a good example. China claims the isolated rocks as their own sovereign territory. But even if we understand that China seeks more security in the international system as its primary goal, the situation concerning the Senkaku/Diaoyu Islands can remain baffling. One has to ponder how such small features could meaningfully contribute to China's security. This is going to be very important for the later stages of the analysis. If a high value can be identified, then competitive behaviour with Japan can be justified. However, if the islands offer no value, then a competitive stance that negatively affects regional security cannot be rationally justified. It is important to note that this is distinct for the cost–benefit calculations in later stages. Those refer to the various options' ability to deliver on these interests. But to be able to conduct such an analysis, obviously those interests have to be identified in the first place.

Understanding what a country seeks to accomplish in a given decision-making situation is also important to determine the basic logic of the choice. Rational choice most often relies on a utility maximizing logic: Achieve the best benefits at the lowest cost. However, such a logic might be inappropriate if a country seeks a specific objective. For example, in a territorial dispute it is possible that the option with the optimal utility is a compromise on joint development: It allows the exploitation of resources with minimal costs, so it would have a favourable cost–benefit ratio. But if a country specifically seeks sovereignty over the disputed territory, then regardless of the favourable ratio, such a compromise would not be acceptable as it doesn't achieve the objective. This will be especially important for feasibility considerations. But again, without having a good foundational understanding of the decision-making situation, one will lack clarity on how to make crucial decisions later in the model.

The other consideration is bounded rationality. One has to understand the decision-maker itself, in this case the state facing the policy decision. This will also be an important foundation that affects all other aspects of the model. Mearsheimer highlights the importance of fear in strategic calculations.[5] In offensive structural realism, fear is described by a negative balance of power. However, as pointed out in Part III, ideational factors also play an important role in whether a country "feels scared". This will affect how much premium it puts on options that can increase its security or how much compromise it is willing to make in favour of options that might not significantly alter the balance of power but deliver other benefits. China's behaviour in territorial disputes can be hardly understood without understanding China as a decision-making agent, including the unique ideational considerations that affect its policymaking. Similarly, Japan's national characteristics will predispose it to favour certain policy decisions over others. If nothing else, deviating from these principles will impose additional political costs that would reduce the attractiveness of certain options. While one can propose to forgo these considerations for normative policy recommendations, in explanatory or anticipatory research they cannot be ignored.

In 2008 joint development negotiations broke down between China and Japan. The main reason was a disagreement on the terms of joint development. Japan sought a halt to the Chunxiao gas field until Japan can invest as a partner. In contrast, China's offer was that Japanese firms will be allowed to invest into the field, but it will fundamentally operate under Chinese oversight and will continue to operate regardless of Japanese investment. The negotiations were fundamentally unsuccessful because understandings of joint development were inherently incompatible. While China emphasized the economic side of joint development, Japan sought it as a political tool to get a recognition that the field accessed disputed resources. Not properly understanding the decision-making agents and what they ultimately sought to accomplish led to the formulation of unsuccessful policy.

Bounded rationality also means confronting the ideational components of security. To effectively pursue through rational choice, one has to understand what a country means by security. Is security defined in purely territorial terms? Is a state's security limited to homeland defence or does it encompass protection of overseas economic interests and sea lines of communications? Is participation in international organizations and the rule-setting process part of the state's security agenda? The answer to each of these questions will vary from state to state. A hegemon will view its security differently from a minor power. Thus, when confronting a similar decision-making situation, they may reach completely different answers.

Overall, a strong rational choice model starts with developing an understanding of the actual decision that needs to be made, including the bounded rationality of the decision-making agent. While it is tempting to use a universalist logic, it will reduce the accuracy of the model. Rationality means purposive behaviour, not necessary adherence to one's own cognitive framework.

Phase 2: Identifying potential options

Once the decision-making situation has been thoroughly understood, the second step is to take stock of the potential foreign policy options a state could adopt to deal with the situation. One can take a simple example: Claiming disputed territory. A state can decide to order its troops to capture the disputed territory. It can engage in coercive diplomacy and threaten the opponent militarily or economically in order to facilitate the handover of the territory. It can seek to negotiate directly or with the assistance of third parties. Or it can decide not to take direct action and adopt a position of observation. This is only a short list of options. The state has a large toolbox available to deal with any particular issue. All of the potential policy avenues have to be identified before the process can progress to reach an accurate answer. If any of the options is overlooked, it will skew the results. Once again one can use an example from the consumer-focused origins of rational choice. If one seeks to purchase a coffee machine, it is prudent to survey the market for all potential competing options. An exhaustive research into the potential options will offer a solid foundation for making the "right" choice when it comes to selecting the final machine. In contrast, if the search has only been conducted shallowly, then it is possible that ideal options are overlooked simply because they are not on one's radar. The International Relations literature helpfully engages with all the potential policy avenues for international conflict resolution. These are discussed later in the following chapters. But to shortly list them here, the key options are as follows:

- Direct use of force (going to war).
- Military coercion (diplomacy of violence).
- Economic/political coercion (geo-economic statecraft and greyzone challenges).
- Bilateral negotiations (compromise).
- Non-binding third-party resolution options (assisted compromise).
- Binding third-party resolution options (court decision).
- Delaying and deterrence.

At the very least one has to be familiar with all of these options in the context of the foreign policy decision. Identifying these options will allow one to proceed to the next stage, that is, feasibility testing, where one can determine what actually can be done.

Phase 3: Feasibility testing

Just because an option exists, it does not mean that it can be realistically pursued. This has been highlighted by Gray in Maxim 21 that states that the impossible is impossible. It is a condition to be accepted, not a challenge for which a solution hasn't yet been found.[6] There is a chance for any plan to succeed. But realistically speaking, it is possible that the chance of success is so low that it is statistically negligible, rendering an option to be pragmatically impossible. It is technically possible that Albania could execute a successful military campaign against the United States. But the chances of it are so low that one can safely discount it when discussing how Albania should formulate policy relative to the United States. It is highly likely that whatever set of options were identified in the previous phase, not all of them will fall within the realm of feasibility. This can be due to insurmountable limitations (e.g., massively unfavourable balance of power) or temporary limitations that nevertheless need to be removed before the policy option becomes viable (e.g., Japan refuses to recognize the Senkaku/Diaoyu Islands dispute, rendering seeking a successful bilateral settlement unfeasible for China).

Feasibility testing focuses on establishing a feasibility criterion for each option and determining whether they reach that in the case of the decision to be made. The feasibility criterion is the minimum conditions that need to be met for the option to be considered going forward. It is distinct from risks that will be discussed in the next phase. Risks refer to the possibility of failure for feasible options. The feasibility criterion in contrast refers to the basic possibility that the option could succeed under ideal conditions. It is possible that a feasible option will have high risks nevertheless, reducing its attractiveness.

How to define the feasibility criterion is a difficult task. Each option has its own set of considerations. Designing a feasibility criterion for going to war will be discussed more in detail below. The feasibility criterion for bilateral negotiations, for example, comes from two main components. On the one hand, one has to take into account the parties' willingness to get to the negotiating table. On the other hand, one has to consider the possibility of a compromise. In the case of the Senkaku/Diaoyu Islands dispute, the feasibility of bilateral negotiations is critically low. Japan refuses to recognize the dispute and maintains that it possesses unchallenged sovereignty over the islands. This will act as a barrier to the any attempt to use bilateral negotiations to settle the dispute short of recognizing full Japanese sovereignty. While this is a temporary block that could be changed by a shift in policy in Japan, the feasibility is further reduced by the nature of the dispute. Much of the dispute focuses on sovereignty, which is indivisible. As long as possession is viewed more important than merely resource extraction, and due to the security importance of the islands it is, a compromise is not possible since two countries cannot maintain sovereign control over the same territory. Based on this one can conclude that in this instance bilateral negotiations do not meet a minimal feasibility to be considered further. One can devise ways to improve the feasibility of specific options, but that is a separate issue from the decision-making situation at hand as it fits better into the context of a multi-level game.

Phase 4: Costs, benefits, and risks

The central idea of rational behaviour in International Relations is to weigh the consequences of different options. This was recognized by Mearsheimer when he argued that states consider the costs and benefits of their actions when adopting policy directions.[7] As part of the rational decision-making process, each feasible option is further studied to determine (a) how they can contribute to the achievement of the overall objective and what additional benefits they can offer, (b) what are associated costs of pursuing that option, and (c) what are the chances of success and how would failure affect the state. This will coalesce into a cost–benefit–risk ratio of the option, which will be the key metric determining its attractiveness for the state.

Benefits can be divided into two major categories. Any policy proposal is expected to contribute to the overarching objective of the state in some manner. To restate, from a realist perspective that overarching goal is security. Thus, whatever issue the state is pursuing there is an expectation that it will somehow contribute to the security of the state. This can be done directly, for example, by affecting the balance of power through the acquisition of strategic territory or new weaponry. But it can also contribute indirectly. As discussed before, Mearsheimer identifies two components of power: actual and latent.[8] Thus, an option that contributes to the economic development of the state also strengthens its latent power, contributing to the state's ability to acquire tangible power in the future. These will be the most significant benefits of any policy direction. But besides how it contributes to the overarching policy objective, options can also deliver other benefits. A good example of these would be shifting a state's reputation in certain directions. If a state acts as a "responsible" member of the international society by adhering to norms of peaceful conflict resolution, it can increase its reputation as a cooperative state and may find it easier in the future to negotiate deals or to be accepted into international organizations. If a state reacts strongly to a perceived threat, a state can build a reputation that it will not pursue appeasement, creating a strong foundation for future deterrence or coercion. Such reputational considerations don't necessarily directly contribute to the achievement of the overall policy objective, but they nevertheless can confer benefits onto the state.

The principle of equivalent exchange argues that for something to be gained, something else must be sacrificed. This principle is clear in International Relations. Any foreign policy option pursued will not only confer benefits but will come with its own set of drawbacks. Some costs, such as material costs, are easy to understand. If a compromise is made to pursue joint development, the resources and profits of the venture have to be shared, creating lower returns than possession would. Other costs are less straightforward. Just as there are reputational benefits to some options, others come with reputational penalties. Aggressive actions or appeasement can create unfavourable reputations. Other countries might perceive the state as a rogue, untrustworthy element or as a pushover as a result of the policy avenues adopted. This will create troubles for future foreign policy formulation. And in some cases, a reputational benefit can also impose certain penalties. For example, settling disputes through certain avenues – for example, through adjudication – creates an expectation that other disputes will be settled similarly. If a country deviates from this expectation, more significant reputational penalties will be applied later on.

Costs and benefits create the foundation of rational choice. Understanding not only what one can gain but what one has to sacrifice and keeping these two forces in balance is at the core of purposive behaviour that one would describe as "rational" in International Relations. However, this assumes that everything goes according to plan. Gray highlights in Maxim 10 that war is always a gamble.[9] This is not only true for war, but all policy avenues. Friction is ever-present and its gradual accumulation can derail even the best-laid plans. This necessitates the introduction of risk into the analysis of various options. The level of risk combined with the risk-averseness of the decision-maker will reduce the attractiveness of an option in addition to its costs.

Once the costs, benefits, and risks of an option have been examined, one can establish the option's cost–benefit–risk ratio. This will be the key rubric that will allow the comparison of each option. Any option where the costs outweigh the benefits can be eliminated from the analysis, as pursuing such an option would be irrational. The ratio in most instances will be based on qualitative research. However, at this stage one might consider quantification: Translating the various factors into numerical values to facilitate comparison. One simple way to do this is a mixture of qualitative and quantitative methods: One can create tiers for each factor with a symbolic numerical value and provide a qualitative analysis of how each component can be categorized. This allows the establishment of a clear cost–benefit–risk ration.

Phase 5: Ranking

The final step of the rational choice process is to compare and contrast the various options based on their cost–benefit–risk rating. If one adopts a utility maximizing logic, then the rational choice will be the option with the most favourable ratio. This can be modified by adding a criterion that certain objective must be met for an option to be adopted – for example, the establishment of sovereign control over disputed territory. In that case the option that achieves that objective and offers the optimal ratio will be adopted as the rational choice. While one can rely on a quantitative methodology as a basis, it is often necessary to engage in qualitative argumentation of why certain options represent more optimal courses than others.

Ranking and adoption can also be affected by bounded rationality. Past experiences will affect the adoption of options and their relative attractiveness.[10] If a decision-maker had success in the past with an option, then it will increase the relative attractiveness of that option. In contrast, if the decision-maker had been unsuccessful in the pursuit of an option, especially if it happens repeatedly, then it will be more hesitant to adopt that option. This is especially important when multiple options possess relatively similar attractiveness to the decision-maker. One example is the submission of the Pulau Batu Puteh dispute to binding conflict resolution through the International Court of Justice (ICJ) by Singapore and Malaysia. Prior to the decision, both parties were eager to adopt adjudication as a way to resolve the deadlocked dispute. However, after the court found in favour of Singapore, Malaysia has shown signs of souring on the process and it has not pursued further cases in front of the ICJ. While states are rational, under the conditions of imperfect information and the resulting unpredictability of risk, past experiences are an important guiding principle for state behaviour, whether one is looking at differentiating friend from foe or at selecting foreign policy options.

Once the comparison and ranking are completed, a policy course emerges as the rational choice. The state now only has to implement that course of action and hope for the best that various elements of friction don't sink its plans. Or, if the research was done in an anticipatory manner, the observer now has a relatively accurate insight into how a state will likely act in the future. While this is a general overview of the rational choice process, the next chapter looks at how each of the key components can be analysed in the context of war itself.

Notes

1 Mearsheimer, J. J. (2014) *Tragedy of Great Power Politics.* Norton: New York.
2 Waltz, K. (2010) *Theory of International Politics*. Waveland Press: Long Grove
3 Angell, N. (1910) *The Great Illusion*. G.P Putman's Sons: New York.
4 Slaughter, A. (2004) *A New World Order*. Princeton University Press: Princeton.
5 Mearsheimer, J. J. (2014) *Tragedy of Great Power Politics.* Norton: New York.
6 Gray, C. S. (2009) *Fighting Talk: Forty Maxims on War, Peace and Strategy*. Potomac Books.
7 Mearsheimer, J. J. (2014) *Tragedy of Great Power Politics.* Norton: New York.
8 Ibid.
9 Gray, C. S. (2009) *Fighting Talk: Forty Maxims on War, Peace and Strategy*. Potomac Books: Dulles.
10 Wiegand, K., and Powell, E. (2011) Past Experience, Quest for the Best Forum, and Peaceful Attempts to Resolve Territorial Disputes. *The Journal of Conflict Resolution*, 55(1), pp. 33–59.

12 The feasibility of war

Whether one can win a war has been an important question for strategic studies literature since the conception of war. War as an activity always incurs significant costs, both in blood and treasure. Thus, a prudent policymaker seeks to avoid engaging in warfare unless there is a reasonable chance of success. From a rational choice perspective, this means feasibility testing and establishing a proper feasibility criterion for the use of force. On the surface this seems simple. The feasibility criterion should be whether one can win or not. However, determining this is a rather complex undertaking. One good example of just how complex determining the outcome of battles, and thus wars, can be is Colonel Dupuy's *Numbers, Predictions and Wars*. The book devises a quantitative model to determine the likely outcome of battles. Some components of it are straightforward: Number of tanks, calibre and range of weaponry, accuracy, etc. However, some components are more esoteric. Dupuy develops numerical indicators for the effects of weather on strategic bombing or how much forest coverage hinders combat operations.[1] This approach has a number of drawbacks. First, much of the information required for accurate modelling might not be available without access to a significant intelligence apparatus. Second, modelling battles is supremely complex even in hindsight. Anticipatory modelling would be even more challenging. And, finally, Dupuy adopts the Western strategic tradition of purposefully ignoring non-material factors, proclaiming the supremacy of technology over men in modern warfare. For the purposes of our analysis, one needs to develop a simpler way of establishing the feasibility of war. The most common approach based on the realist tradition is to focus on the balance of power between competing states. Unfortunately, balance of power analysis often goes to the extreme in the opposite direction. If quantitative modelling such as Dupuy's is too complex, balance of power analysis is often too simplistic. One cannot simply take the total military power of one country and then compare it with the total military power of another country and gain an accurate understanding of the outcome of a potential war. This ignores key factors such as the political nature of war, the rational correlation of objectives and efforts, and the fact that not every war brings the total power of the nation into the fight. This has been recognized since Clausewitz.[2] One needs a more nuanced understanding and as ever so often it is the solution that lies in developing a deeper understanding of the context of use of force. To understand whether a country can win a conflict, and thus pass the feasibility criterion, one needs to understand what conflict the country will be fighting and then determine whether it possesses the capabilities to do so.

The suitability of the objective

The use of force is an important component of the state's foreign policy toolbox. But it is not suited to all issues that a state faces in the international system. When examining the feasibility of war, the first question one should ask is whether the desired goal can be accomplished through war in the first place. War is a suitable tool for objectives that require seizing and holding objectives or if a state seeks to compel another to perform certain actions. War is less suitable for goals that require a change in the hearts and minds of the local population. Most societies tend to harbour negative feelings if one visits the destruction of war on their lands. This is not to say that war cannot be part of a strategy of transforming societies. But it cannot be the only component, merely a tool to overcome initial resistance and/or remove barriers. This sentiment is echoed by Gray in Maxim 3: It is more difficult to make peace than war.[3] The ability to translate military success into post-war political victory is a daunting task.

The United States has received a stark reminder of this through their military endeavours in Iraq and Afghanistan. The goal of the Bush Administration was not to seize strategic territory, to destroy elements within the territory, or to compel these states to surrender elements sought by them. The overall objective of President Bush was to transform these societies into democratic countries that would act as responsible members of the international community. Achieving this through American joint-operations warfare proved to be impossible. Iraq continues to struggle with post-war democratization as it confronts Islamic State of Iraq and Syria (ISIS) and other legacies of a sudden power vacuum caused by US efforts in the country. In Afghanistan, the United States has faced a decade-long insurgency that it now seeks to end by negotiating a peace agreement with the Taliban in order to finally secure an exit from the country. In both of these instances, one can see that the overall objective of the conflict was not suitable for the tool selected. Democratization in most cases is not a suitable objective for war. The US experience has shown that through warfare alone, one can install a supposedly democratic regime, but one cannot solve the deeper underlying political issues that would need to be resolved to create a functioning democracy. One also cannot make the population develop a positive association through use of force alone. The American failure in Iraq and Afghanistan is in stark contrast to their success in Japan after World War II. While it was the military stage of the conflict that allowed the Americans to reshape the country, it was actual post-war political engagement that transformed Japan from an enemy into a friend.

In contrast, territorial disputes offer a more suitable environment for use of force. A disputed territory is an objective that can be seized and held, falling under Schelling's definition for the primary use of force.[4] Through dispatching troops one can establish *de facto* sovereignty over a territory and through armed resistance one can force an opponent to accept the status quo and cease the military phase of the conflict. It does not require the deep internal changes in the opponent that, for example, successful democratization or other identity changes need. An opponent does not need to like the situation or even accept the outcome; it simply needs to accept that it has no available course of action to seek a recourse. Schelling argues that:

> With enough military force a country may not need to bargain. Somethings a country wants, it can take, and some things it has it can keep, by sheer strength, skill and ingenuity. It can do this *forcibly*, accommodating only to opposing strength, skill and ingenuity and without trying to appeal to an enemy's wishes. Forcibly a country can repel and expel, penetrate and occupy, seize, exterminate, disarm and disable, confine, deny access, and directly frustrate intrusion or attack. It can, that is, if it has enough strength. "Enough" depends on how much an opponent has. [5]

Any political objective that falls within these parameters can be feasibly pursued through war. Objectives that are beyond these set of definitions, for example, identity change, are not a suitable objective that can be pursued through war as a foreign policy tool. Trying to force a square peg into a round hole will produce suboptimal results, and in the case of war a terrible cost. Trying to reconstitute an opponent through a military strategy resulted in the United States sacrificing thousands of lives, contributing to the deaths of numerous civilians, creating an ideological foundation for new extremist groups, and in return Afghanistan or Iraq is hardly closer to peace or being an effectively functioning member of the US-led order than it was before. If anything, Iraq has been moving closer to Iran and the Taliban has experienced a resurgence in Afghanistan. Understanding the suitability of the objective for the tool employed is going to be fundamental to the feasibility of war.

Understanding the type of conflict: Limited war versus total war

One of the key principles of Clausewitz is that the effort exerted in war needs to be proportional to the political objective sought.[6] This is connected to the basic idea of rational behaviour described earlier in the chapter that is strongly linked with International Relations theories such as realism. At the same time, Clausewitz also notes the difficulty of controlling escalation in war.[7] This sentiment is echoed by Gray in Maxim 7, which argues that albeit reason should rule in war, passion and chance threaten to overrule it at every turn.[8]

One can illustrate this through a simple example. We want to detach a mineral-rich territory from our neighbour. We determined that the territory is worth sacrificing 5,000 out of our 100,000 troops maximum (5%). Any further sacrifice would render the investment irrational due to the costs outweighing the benefits. Ideally, we would lose far less soldiers, but that is a separate question. Based on this math we begin the operation. A military defeat has been dealt to the local enemy forces and the objective is achieved within acceptable costs. However, the enemy is not ceasing combat operations. Rather it is ramping up military activity, mobilizing its total national military power to resist. In a system governed by imperfect information and worst-case scenario thinking, it fears that our initial victory has opened up the way for further military activity. Now we have a total war at our hands, where the survival of the nation is at stake, pushing the conflict beyond its rational limits.

Escalation in war is hard to contain because any victory shifts the balance of power, making the opponent more insecure and more fearful of its own survival, leading to escalatory behaviour. This does not mean that a limited conflict cannot be successfully pursued. Even Clausewitz, despite his original opposition to the concept, came around to this point.[9] But it means that it is not guaranteed that just because one wants a conflict to be limited it will remain limited. Reaching back to Maxim 16, the enemy too has a vote in the affairs of war.[10] One needs to critically engage with the idea of limited war in any conflict, focusing on two main points. First, how much does the opponent value of the object sought? Just because one attaches limited value to the objective, it is not guaranteed that the opponent won't fight till the last man to keep it. Second, is there a larger context that makes a limited war impossible? For example, if capturing territory will remove all defensive lines between one's own, now victorious, forces and the opponent's capital, then they are unlikely to accept the conflict as a limited war. Similarly, if, for example, in the case of a limited maritime conflict one strikes the home territory of the opponent, it is similarly expected to lead to escalation. This was one of the main concerns surrounding the implementation of the Air-Sea Battle concept.

War is essentially fire. With proper care, it can be an effective tool. However, it always inherently seeks to break free from its constraints. If not handled carefully, a single flame can cause a forest fire. Thus, the feasibility of war as an option significantly comes down whether it is possible to control the flames of war in the particular context of one's objective to a degree that the benefits of the endeavour remain ahead of its costs.

Julian Corbett, the English naval strategist, offers valuable insights into keeping conflicts limited. He outlines two principles to keep a conflict limited. First, the object of the conflict has to be of limited political value for both sides. Second, the theatre has to be able to be isolated.[11] The first point should be self-evident at this point. But the second point is worth some consideration. The idea of strategic isolation means that a theatre can be set up in a way that whatever happens within it won't spill over outside the theatre. One easy example is to imagine that two nations are fighting to gain control over an island. Both can use their navy to isolate the island, meaning, regardless of who wins, the victor can't directly threaten the home territory of the other.

In such a scenario a conflict can remain a limited contest. This is important, as some of the most significant regional security hotspots are disputes over remote islands, in which case such isolation is feasible. And if such isolation is feasible, the possibility of a limited war becomes more threatening. These islands are not worth a total war, but a limited military contest is not out of the question, as one can see in the cases of the 1974 Paracel Islands and 1988 Spratly Islands conflicts between China and Vietnam. Rational behaviour requires the careful weighing of the possibility of keeping a conflict limited to assess the feasibility of use of force. Most objectives are not worth a total war, especially when one takes into account nuclear armed major powers inhabiting the system. Although one could argue that nuclear arms themselves can contribute to keeping a conflict limited: If total war means the extinction of the human race, then both parties are more eager to keep a conflict limited (as was explored in Chapter 4). This is especially so as if an opponent seeks to capitalize on its earlier victory to launch a larger invasion of one's home territory, a devastating second-strike ability is always kept in reserve. But this creates a game of "chicken", in which scenario it is possible that success has to be abandoned in order to avoid total nuclear annihilation.

One additional consideration to this is reputation. One theme that keeps reoccurring is that under the conditions of imperfect information perceptions are paramount. Under such conditions, reputation is a useful indicator. If a state has a reputation to escalate conflicts, one should be more hesitant to pursue a limited war. In contrast, if an opponent has a reputation for more rational behaviour, one can be more confident to attempt it. Past actions are not a guarantee for future behaviour. But it is nevertheless an important indicator

Understanding the possibility and the attached difficulties of keeping a conflict limited is essential to determining feasibility. If the probability of escalation is too high, then feasibility is diminished, and one is better served looking at other tools instead.

The balance of power

Assuming that one has developed a deep understanding of the conflict one has to make a decision about, and that they allow for the use of force, then the only question left is whether one can actually successfully pursue war to achieve one's objectives. To determine this, one can look back to the quote from Schelling above. "It can, that is, if it has enough strength. 'Enough' depends on how much an opponent has".[12] The fundamental metric for the feasibility of war, as to many elements of international relations, is the balance of power between competing states. At this point, it is important to note that one is looking at whether a country has a reasonable chance of success, not whether it is going to win the conflict. The focus is probability not certainty. Determining the basic feasibility of military action is a straightforward equitation: One takes the available military power one can rationally allocate to the pursuit of the objective and subtracts the military power the opponent has to resist. If the result is positive, one has a favourable balance of power. If the result is negative, then one faces an unfavourable balance of power and military action might be beyond feasibility. However, in practice, determining the balance is more complicated.

Determining the feasibility of war is often based on qualitative argumentation rather than neat quantitative analysis. One can approach the feasibility criterion from a negative perspective. The question becomes not whether the balance of power is favourable to the country, but rather one focuses on whether there is an obstacle in the form of a negative balance of power that would "diminish" the feasibility of war. On the one hand, this is less precise than a positive approach. On the other, it allows the definition of a clear and utilitarian feasibility criterion, since one is less concerned about the precise balance and more focusing on whether there is any reason why war would not be feasible. Once feasibility is established, it is still possible that later on war will not be selected as the optimal course of action due to high risks based on a lack of advantage in the balance of power. Even with the simplification of the feasibility criterion, analysing the balance of power requires critical care.

The first point to note is the difference between total and situational balance of power. Total balance of power is the sum of all military power possessed by a country. Strategic analysis often compares this number between countries. This can be useful if one is analysing the distribution of power within the international system on a systemic level. However, when it comes to analysing the feasibility of war, it is a largely useless indicator. This is indicated by Clausewitz when pointing out that even in a total war context the total military might of a country is rarely brought into play, remaining largely an abstract concept.[13] Instead what matters is the local or situational balance of power. This builds on an understanding of the particular context of the conflict. For example, in a conflict predominantly fought at sea, the number of ground troops and tanks is largely irrelevant. Similarly, even the best navy will not help fighting in a landlocked theatre. Air superiority fighters do not affect the balance when fighting insurgent troops dispersed within the local population. The number of battle tanks is of no concern if one does not have the amphibious capability to transport them to a distant theatre. What the context of the actual conflict is determines the type of military power needed and thus the balance of power specific to the war at hand.

Another fact to consider is that material factors are clear indicators of battlefield performance. Western strategic tradition often displays a certain disdain for the human element. However, one ignores the importance of human motivation at their own peril. Other strategic traditions, such as Mao's People's War, put morale and other nontangible factors into the forefront of strategic planning. It is viewed as a way to make up for material weaknesses when facing technocratic opponents such as the United States. Unfortunately, the balance of power is affected by these hard-to-judge factors such as morale, effectiveness of chain of command, initiative and expertise of field commanders, and so on and so forth. Most of these are inherently unknowable. For modelling purposes, one can black-box these factors and accept that one's model will suffer a hit to its accuracy. Again, the point is not to achieve certainty, but to pursue reasonable probability. One potential way to incorporate these considerations into the model without the intelligence apparatus of a state is once again to adopt a negative approach. Instead of focusing on how they contribute to feasibility, one can shift their focus to see whether there is any major issue that would negatively affect feasibility. While one might not be able to gather information of the battle prowess of specific commanders, issues such as endemic corruption, lack of qualified manpower, or poor training practices can be identified through open-source strategic literature. If identified, one can draw the reasonable conclusion that the feasibility of war is reduced, unless there are corresponding weaknesses on the other side.

How this looks like in the real world can be demonstrated by analysing the balance of power between China and Japan. If one looks at the total balance of power between the two countries, the picture is obvious. China is arguably one of the top three military powers on the planet next to the United States and Russia. In contrast, while Japan possesses formidable military power, it is only a major power. Looking at traditional indicators used by offensive structural realism, we can get a basic comparison, as shown in Table 12.1.

This provides a basic overview of the balance of power, seemingly favouring China, quite significantly. From an offensive structural realist perspective China is more powerful than Japan. This is correct on the systemic level. Overall, China has more power in the international system than Japan. That is not a controversial notion. However, this is largely an abstract concept. More relevant to the discussion at hand is a different question: Can China win a war against Japan in the context of the Senkaku/Diaoyu Islands dispute? Relying on purely the indicators discussed above, one would conclude with a firm "yes". But the reality is not so simple (Table 12.2).

The particular context of the dispute means that it would be primarily a conflict fought at sea and in the air as the Senkaku/Diaoyu Islands do not offer space for large land operations. Looking at the pure numbers, one can already see a reduction in the supposed advantage China has in the balance of power.

Table 12.1 Balance of power between China and Japan

	China	*Japan*
Troop Numbers (Active)	~ 2.1 million	~ 250 000
Military Budget (Active)	~ 237 million USD	~ 49 million USD
Population (Latent)	~ 1.4 billion	~ 126 million
GDP (Latent)	~ 13.6 trillion USD	~ 4.9 trillion USD

Source: globalfirepower.com, except GDP, sourced from World Bank.

Table 12.2 Context-specific balance of power

	China	*Japan*
Number of Ships Total	777	155
Number of Destroyers	36	40
Number of Submarines	74	20
Number of Combat Aircraft	1232	279

Source: globalfirepower.com, Balazs Szanto (2017). *China and the Senkaku / Diaoyu Islands Dispute*. (Routledge: London).

Importantly, Japan has an advantage in the number of surface combatants, the primary vehicles for the combat scenario. While China maintains a large frigate fleet, these are significantly less capable than destroyers due to smaller size and less weaponry.

But if one actually looks beyond the numbers, the picture becomes more interesting. Japan primarily relies on battle-tested US technology. In contrast, while China has a numerical advantage, it primarily comes from keeping older, less capable models in service. The Ming-class submarines, while heavily updated, are based on old Cold War hulls. China's air force is also made up of a large number of indigenously produced fighters and the naval aviation arm is prone to receive handed down aircrafts from the PLA Air Force. A large chunk of China's destroyers also comes from various experimental classes whose design was abandoned for various reasons, albeit recent years saw a mass production of Type 052D-class destroyers. But overall, there are questions about the effectiveness of China's weapon systems relative to that of Japan's. This significantly reduces the potential advantage China would seem to have based on numbers alone. This is compounded by the fact that the conflict would likely be a limited war. In this context the total military power of the combatants is less relevant compared to their unit-to-unit combat capabilities. It is unlikely that the bulk of either the Japanese or Chinese forces would be deployed. Thus, China's numerical advantage is less relevant compared to potential technological and other weaknesses. Other weaknesses include continued issues with corruption, including when it comes to promotions, training and equipment issues, especially for theatre air defences and anti-submarine warfare operations, and a rigid command structure with a lack of initiative on the lower levels.[14]

Once the case is studied in more detail, it becomes clear that the balance of power between China and Japan in this case is more equal than the numbers alone would suggest. While military action for China is certainly feasible, it would be a significant gamble, especially if Japan's defence ties with the United States are taken into account. Connecting back to the original discussion on the feasibility criterion, there is not a negative condition that would make war impossible, thus the feasibility criterion is achieved. But what seems to be an obvious answer from a high-level perspective becomes much more muddled once examined in the context of a particular conflict. And this is where the dangers of perception in a system governed by imperfect information can be tangible. China has been projecting an image of strength for decades and thus it is easy to overvalue its military might, especially as the PLA has not engaged in a major conflict in decades and thus we lack battlefield experience to test its actual military readiness. In contrast, Japan has traditionally kept a low security profile and as such one tends to undervalue the fighting power of the Japanese Self-Defence Forces. This creates a perception of a gap between the two countries that appears much bigger than it really is. Much of China's military might lies in PLA ground troops, which cannot be brought into a conflict in the maritime realm. Understanding the particulars of how military power might be used to achieve an objective will be fundamentally important to see whether feasibility can be reached as it provides a more nuanced and realistic analysis relative to a focus on basic total balance of power comparisons.

Establishing feasibility is only the first step of the rational choice comparison. Just because an option is feasible it does not make it the optimal course of action. The next chapter looks at the benefits, costs, and risks of the decision to go to war, which can lead to the rational decision to go to war.

Notes

1 Dupuy, T. N. (1979) *Numbers, Predictions and War: Using History to Evaluate Combat Factors and Predict the Outcome of Battles*. Macdonald and Jane's: London.
2 Clausewitz, C. (1874) *On War* (Translation by J.J. Graham). Guttenberg.
3 Gray, C. S. (2009) *Fighting Talk: Forty Maxims on War, Peace and Strategy*. Potomac Books.
4 Schelling, T. (2008) *Arms and Influence*. Yale University Press: New Haven.
5 Ibid., p. 1.
6 Clausewitz, C. (1874) *On War* (Translation by J.J. Graham). Guttenberg.
7 Ibid.
8 Gray, C. S. (2009) *Fighting Talk: Forty Maxims on War, Peace and Strategy*. Potomac Books.
9 Handel, M. I. (2001) *Masters of War*, 3rd edn. Frank Cass: London.
10 Gray, C. S. (2009) *Fighting Talk: Forty Maxims on War, Peace and Strategy*. Potomac Books.
11 Corbett, J. (1911) *Some Principles of Maritime Strategy*. Guttenberg Online.
12 Schelling, T. (2008) *Arms and Influence*. Yale University Press: New Haven, CT, p. 1.
13 Clausewitz, C. (1874) *On War* (Translation by J.J. Graham). Guttenberg.
14 Chase, M. S., Engstrom, J., Tai M. C., Gunness, K. A., Harold, S. W., Puska S., and Berkowitz, S. K. (2015) "China's Incomplete Military Transformation: Assessing the Weaknesses of the People's Liberation Army (PLA)." RAND National Security Division. <www.rand.org/content/dam/rand/pubs/research_reports/RR800/RR893/RAND_RR893.pdf> (accessed 11 April 2016).

13 The costs, benefits, and risks of war

Once the determination has been made that war is a feasible option, one needs to still examine whether it is the best choice under the circumstances. Just because something is possible, it does not mean that it is automatically an attractive option. Rational choice modelling usually focuses on identifying the optimal or best course of action from a set of feasible options. This chapter critically examines the costs, benefits, and risks of war, which is essential for working out its relative attractiveness compared to competing options that are discussed in the next chapter. Working out an assessment framework for these factors can be complicated, especially when it comes to translating them into a contrastable ratio, since it requires the comparison of dissimilar factors. Once again, a deep understanding of the decision-making agent and its bounded rationality framework is needed to judge how a particular option serves their interests.

Understanding the benefits of war

Contextualizing the benefits of war will primarily focus on its ability to solve difficult problems and accomplish objectives that would be impossible or at least very unlikely to achieve through other approaches. War, or the direct use of force, can accomplish a variety of objectives, and does so without the need to bargain with the opponent. How this looks in practice is that if one desires a particular territory, one can send in the army to occupy the territory forcibly. This would be the most standard example of war used to accomplish an objective. However, the direct use of force is not limited to seizing territory. If one is dissatisfied with the policies of a nation, one can lead a "punitive expedition" against the country, in which case the goal is not to seize the territory but to force a political change or to express dissatisfaction through arms. An example of this would be the 1874 Japanese punitive expedition against Formosa (Taiwan) for the killing of shipwrecked Japanese sailors or the 1916–1917 "Pancho Villa Expedition" by the United States against Mexico. Similarly, when the People's Republic of China (PRC) pursued nuclear weapons, at certain points both the United States and the USSR proposed the use of nuclear arms of their own to put an end to it. In

1981, Israel launched a pre-emptive airstrike to prevent the construction of a nuclear reactor in Iraq. At the height of the Iran hostage crisis (1979–1981), the United States launched a military operation to free the hostages and end the stalemate with the new revolutionary government in Iran. In 1983 the United States invaded Grenada, alarmed by the construction of an airport for potential military use by the Soviets and responding to a growing political crisis. As these examples show, the use of force can be applied beyond securing territory, in order to accomplish a variety of objectives, as long as the objective can be accomplished through high explosive ordinances.

The primary benefit of going to war is that one can pursue an objective without the need to compromise with an opponent. As long as one possesses sufficient power to accomplish the objective, the opponent's objectives are irrelevant to the calculation, as all they can do is to resist through arms. This is not to say that bargaining cannot be part of war. Especially if one wants to achieve a formalized peace, then the enemy's will has to be diminished. But this is not required for war. As long as one possesses sufficient military power and determination, one can hold on to the objective and there is not much an opponent can do other than to engage in counterattacks. Eventually either one or the enemy will reach the rational limits of persistence, leading to de-escalation if not formal peace or acceptance of the status quo. North Korea and South Korea have never reached a formal peace agreement after the Korean War. For over 60 years the two countries have been formally at war, yet hostilities remained sporadic, both holding onto their territories through armed deterrence. An even more relevant example is the PRC and Taiwan. While Taiwan has some legal protections, it is not recognized by most countries as a state. And the Chinese civil war has not come to a formalized end, nor is China any more willing to accept what it considers to be a renegade government to maintain formal control over Chinese-claimed territory. Yet, rational considerations have frozen tensions. India and Pakistan have been feuding over Kashmir, both holding onto parts they can secure militarily; yet, despite occasional escalation, armed confrontation is not ever-present. In the end, while it is preferable to have a formal end to a conflict with both parties accepting the status quo, it is not inherently necessary for the successful conclusion of a military campaign, especially if the goal was not to accomplish territorial gains but to pursue one of the many other examples mentioned above.

At this point, it is important to note the distinction between *de facto* and *de jure* in the International Relations vernacular. *De jure* refers to concepts that have basis in international law. For example, *de jure* sovereignty is accomplished through discovery or prescription, and it is considered to be legal under international law. As such, it enjoys certain protections within the international community. If *de jure* sovereignty is violated, the victim can usually petition the international community, for example, through the United Nations, for redress. In contrast, *de facto* has basis in fact, but not in law. If a country invades another country and establishes control over one of its provinces, then it established *de facto* sovereignty over the area. Since the use of force is contrary to international law, such possession, while recognized as a factual reality, is not legally valid. *De facto* possession does not entitle a country to the same legal protections as *de jure* sovereignty. If anything, since most *de facto* acts have their roots in actions that are viewed as contrary to international law, the international community is likely to seek to reverse them. Thus, they can only be maintained through continuous exercise of force. By definition, *de facto* possession is weaker than *de jure*. However, it can be valuable if there is no *de jure* way to accomplish an objective. This connects back to the point above that war can resolve issues that proved to be resistant to other forms of resolution options, even if the resolution might be contrary to international law.

One might pursue a *de facto* resolution if a *de jure* resolution would require an unacceptable compromise or if an opponent is unwilling to negotiate about the issue. Taking the 1981 air raid as an example, if Iraq refuses to cease construction of a nuclear power plant and Israel determines that preventing its completion is a high-value interest, then Israel can overcome Iraq's refusal to negotiate through directly targeting the installation. No negotiation is necessary – it is simply the force to overcome whatever defences Iraq can muster to prevent such an outcome. The Paracel Islands were equally divided between South Vietnam and the PRC prior to 1974, an arrangement neither party was happy with. In 1974, China took the other half of the islands by force and has been holding onto them ever since. While its possession could be legally challenged, it maintains *de facto* administrative control, just as it does over the Mischief Reef or the Scarborough Shoal. This ability to disregard a non-cooperative or obstructionist opponent is the primary attraction of use of force.

Determining whether *de facto* results are sufficient is going to be a key question for the attractiveness of use of force. In a territorial dispute, if one wants to construct a military facility, *de facto* control is enough. If one wants to establish an exclusive economic zone (EEZ), it is not, as EEZs are rooted in international law, which would view possession of the territory as illegal. Connected to this is the idea of whether one needs the enemy to accept the outcome. If the goal is to blow up a nuclear installation, retarding progress, then war can be an attractive option. If one wants an opponent to give up on the project, then it might not be the case. Nevertheless, while the latter would be ideal, sometimes one has to settle for the former in the case of a non-cooperative opponent. War essentially functions as a last-resort option. Through its ability to disregard the will of the enemy, it can accomplish important tasks if all other avenues failed or would be irrationally costly.

Another benefit of war is that it can be used to affect one's reputation in the international system in desirable ways. Reputation for resolve is a highly valuable asset to possess as it underpins both coercion and deterrence. At its core, both coercion and deterrence (which can be conceived as defensive coercion) are based on threats: A threat is issued if the opponent doesn't behave in a certain way, then there will be significant, in this case military, consequences. Fundamentally both are a function of the balance of power. A state can issue threats based on its capacity to carry out successful punitive action relative to the defensive or retaliatory capabilities of the opponent. However, the balance of power on its own is not sufficient to explain successful coercion. A threat needs to be credible for it to be effective. Just because a state has the power to engage in punitive action, it does not mean it also has the willingness to do so. States routinely draw lines in the sand in order to coerce their opponents, but they often do not follow through with these threats. The Obama Administration had its red lines concerning the civil war in Syria, but it did not intervene due to larger geopolitical considerations, despite evidence that these red lines have been crossed. It is prudent for any state to be sceptical about the credibility of any threat before deciding to abide by it. This raises the question of how one can determine whether a threat is credible, especially as the balance of power is often not a useful indicator. Once again, reputation can be a good starting point. If a state has a reputation to follow through its threats, then one should proceed with caution. This reputation is built through using force to retaliate for unacceptable behaviour (assuming that the threat has been a military one and not an economic one such as sanctions). Selecting use of force as a foreign policy tool can help build this reputation of resolve, which can make future attempts at coercion or deterrence more successful.

Overall, the two main benefits of use of force (or going to war) are the following: (a) It can deliver objectives that proved to be resistant to other policy avenues and (b) it can build the state's reputation to facilitate credible coercion. However, while there are obvious benefits of going to war, it also has significant drawbacks, which will be discussed next. The decision whether one should go to war will be underpinned by whether the costs outweigh these benefits.

Understanding the costs of war

War is a supremely expensive undertaking. War as a large-scale organized violence results in the tragedy of death and destruction. Conflicts such as World War I and World War II, the Vietnam War, or the American Civil War have effects comparable to plagues or other acts of god. The sacrifice required in blood and treasure to accomplish one's interests is massive. Modern wars have sought to reduce the death toll and shrinking the scope of the conflict seeks to lower the sacrifice necessary. In the shadow of nuclear weapons humanity can hardly afford the periodic burning of the world observed in the first half of the 20th century. Yet, even fighting a modern, limited, surgical-strike-focused conflict is supremely expensive. A Harpoon anti-ship cruise missile block costs around 1.2 million US dollars. While the block includes auxiliary components that can be reused, firing a single shot still requires significant material investment. And while the overall lethality of war has been reduced, from a purely economic perspective, training a modern, technologically savvy soldier requires significantly more investment than simply giving rifles to peasants, and thus their loss is correspondingly damaging. For war to be rational, this sacrifice in blood and treasure has to be lower than what is gained. Otherwise, one is simply cutting off the nose to spite their face.

But beyond simply how much the equipment and soldiers cost that the war burns through, one also has to consider that these losses will take time to recover. Achieving an objective through war, only to suffer a significant loss in the balance of power and then to be overtaken by another competitor would be a pyrrhic victory. This was recognized by Sun Tzu when arguing that one should not commit to protracted warfare or seek to besiege fortified cities as these will exhaust the army and the state, leaving it vulnerable to attacks from others.[1] This is an additional reason why the total military power of a nation is a largely irrelevant concept, unless one considers the scenario of a classical total war. Every nation has multiple security commitments and thus can hardly afford to commit most of its military power to a single conflict. Ultimately, at the end of a military conflict one has to have enough power left to at least carry out the basic mission of national defence while it recovers from the losses. It is very possible that, even if the immediate material investments would be worth it, medium-term loss of mission readiness makes a war not worth it.

Furthermore, the sacrifice in blood and treasure doesn't necessarily stop the minute the objective is accomplished: the war can continue beyond that point As discussed above, *de facto* results need constant military maintenance. In the case of a disputed territory, this means occupation, which can be more costly than the actual combat phase to acquire the territory. The cost of defending the gains or resisting a retaliation from the opponent after the actual objective has been achieved might far exceed that of the initial phase of the war. It is important to remember that once unleashed, it is hard to contain war. There is no force that mandates that enemy ceases combat as soon one's objectives have been accomplished. Even from a moral perspective, just war theory recognizes the right of the attacked not only to redress the aggression but to engage in a punitive campaign against the aggressor.[2] Thus, one is likely to experience losses even after the objective is accomplished until the enemy's will to fight is exhausted.

One example that can illustrate the insidious nature of material costs is the US involvement in Iraq and Afghanistan. While the initial combat phase was successful, the ensuing military occupation proved to be costly, especially as sustained occupation was necessary in the face of an enemy unwilling to accept defeat and resorting to insurgency tactics. The material costs were significantly higher over the decade-long occupation than during the initial war. Also, the US military was slowly transforming to accommodate mission demands. At the height of the insurgency, the US experience increased demands for ground troops, mine-resistant vehicles, and drones. While these are valuable for a counter-insurgency force, they are of limited use against a near-tier competitor. Such a conflict would require submarines, air superiority fighters, and guided missile destroyers, which in turn are not of much use against insurgents. For a country with limited means, such demands might mean forgoing acquiring other equipment, affecting the overall mission readiness.

This raises the question of how much actual material investment one needs to make into any particular conflict. This primarily depends on how much resistance an enemy can or is willing to put up. Essentially just as one would conduct a cost–benefit analysis on how much an objective is worth to themselves, one also needs to conduct a similar analysis to determine how much the opponent values not giving it up. This will determine how much resistance the opponent is expected to put up, based on which one can calculate the force needed to overcome it. This follows the same methodology one would use for their own calculations but requires a sufficiently deep understanding of the opponent to be accurate. Ideally one has to consider a favourable balance of power to provide a margin of error that can account for acts of chance and friction.

Beyond the immediate and extended material costs, one has to take into account that, if one goes to war, then one violates international law. The use of force, in most cases, is clearly against the accepted norms of the international system (Article 2.4 of the United Nations Convention). Thus, choosing to go to war is likely to draw a punitive response from the larger international community. A recent example would be the Russian annexation of Crimea. Moscow was argued to have supported local militia groups to facilitate its *de facto* acquisition of the Crimean Peninsula. While based on a popular vote, the Russian acquisition is largely considered to be against international law. Combined with Russian military activities in Ukraine, the international community saw it justified to impose sanctions against the Russian Federation. Both the European Union and the United States have initiated targeted economic sanctions to cripple the Russian economy and force Moscow to reconsider its actions. North Korea has been under biting economic sanctions for decades due to its development of nuclear weapons. Similarly, Iran has been under strict sanctions, primarily from the United States, for its own nuclear and regional ambitions. The international community often displays a willingness to punish states that do not observe international norms. The most extreme form of this is a full military intervention. This can take the form of collective self-defence or an international intervention to separate combatants and enforce ceasefires. Article 51 of the UN Convention provides the right to any state to intervene on behalf of a victim of aggression. The US intervened on behalf of Kuwait during the Gulf War. China has intervened on behalf of North Korea against UN forces during the Korean War. When considering the costs of war, one has to take into account the response of the larger international community and its capacity for a wide range of punitive actions ranging from economic and political sanctions to a full-scale military intervention.

Determining the international response can be difficult as the international community has a less than consistent record on the enforcement of norms with issues often being wrapped up in the political interests of major powers. But one can make some observations nevertheless. The collective response of the international community is largely directed through the United Nations. The permanent members of the UN Security Council hold veto powers, effectively possessing the ability to block most collective action. This can shield them or their allies from a wide range of collective action. But states still can take action outside the UN framework. The sanctions against Russia were coordinated by the United States and the European Union, but they were not within the framework of an international governmental organization (IGO). The degree of negative consequences often comes down to the power held by that state in the international system and their ability to influence other states. Powerful states can deter a lot of international interference. For example, while one can specifically target the Russian energy sector for sanctions due to Moscow's high dependence on it, targeted sanctions would be difficult to implement against China. China has no key sector that could be targeted similarly and due to its high level of integration into the world economy sanctions would impose significant costs on the global economy as a whole. When the United States ignored the International Court of Justice's ruling on its support for Contras in Nicaragua, there was very little any state could do to change US behaviour. Similarly, when the United States invaded Iraq in 2003, it did so without a clear UN authorization and against the objections of several major powers. Yet, no significant consequences befell on the United States. Isolated, self-reliant states, strong states, or states allied to strong states can weather international consequences. As an example, China is often accused of sanction busting on behalf of North Korea.

Ultimately, violating international norms will impose additional costs. But the extent of these costs depends on the balance of power in the international system. At the end of the day, despite the existence of the United Nations, the system is still anarchic. There is no government overseeing the international community that could provide enforcement similar to the domestic environment. "Bad behaviour", if managed carefully, can be profitable as the international community possesses limited means of recourse to control the behaviour of sovereign states. And it does so by design. No major power would consent to membership if their behaviour would be critically constrained.

A final consideration is the effects of use of force on reputation. As discussed above, it can be useful for building a reputation for resolve. But the use of force also comes with reputational penalties. Having a reputation for use of force, and thus a reputation for not playing by international norms, will exert a negative influence. In an anarchic system cooperation is only possible based on some level of trust. If a country gains a reputation that it is willing to use force, other nations will increasingly regard it as a threat. This behaviour is at the centre of defensive structural realism. Waltz argued that conquest is often self-defeating due to other states forming a balancing coalition to defeat the common threat.[3] China has a first-hand experience of this. Due to its growing military might and increasingly assertive behaviour, the PRC has been subject to the so-called "China threat" theories, growing regional opinions increasingly characterising the country as a regional security problem. Growing threat perceptions negatively affected China's ability to effectively deal with some nations through diplomatic channels, prompting Hu Jintao to pursue China "good neighbour" policies to reduce such threat perceptions. Essentially, if a country is mistrusted and feared, it will find it more difficult to engage in policy directions reliant on more subtle persuasion. In extreme cases the international community might seek to isolate them altogether. Many regimes labelled as belligerent by the international community have suffered this fate, whether one is talking about North Korea, Iran, or Cuba. Thus, when discussing the costs of use of force, one has to take into account such reputational costs. The extent of the damage can depend on many factors, such as the country's influence in the international system or whether it managed to provide adequate justification for its use of force. But repeated patterns of aggression will lead to an unfavourable reputation. The United States occupies a hegemonic position in the system and it maintains close alliances with several other major powers. However, its willingness to resort to the use of force, whether during Cold War proxy conflicts, post-Cold War interventions, or the War on Terror, creates a reputation even among allies that the United States can be a volatile actor. The perception that the United States would bomb first and sort out the details later has become part of popular culture. This can make other states wary of US policies and countries such as China regard it with heightened threat perceptions.

Overall, war can be viewed as a double-edged sword. While it can confer significant benefits, especially if other tools of statecraft have failed, it always imposes a significant cost. The concluding maxim of Gray has been "tragedy happens" for a reason.[4] The occurrence of war is always tragic and it will impose significant costs on the participants. However, occasionally such sacrifices are necessary. The tragic nature of war alone is not sufficient to discard it from the international system. However, before one embarks on a course of war, it is important to be clear on whether the benefits outweigh the costs, which is a tall order in most cases. War is a tool of last resort because its use will always require a significant sacrifice in blood and treasure, and that assumes that the conflict is won. The final component of understanding the attractiveness of an option is to examine its risks because things hardly go according to the plan in international politics.

Understanding the risks of war

To quote Conrad Brean from the film *Wag the Dog*, "all combat takes place at night, in the rain, and at the junction of four map segments"[5]. War is a complex activity, and if things can go wrong, they will. Troops will get lost. Two warships will accidentally ram into each other. A sandstorm rolls in blocking air cover. Or if one is part of the Russian fleet heading to the Far

East during the Russo-Japanese war, then winter cloths will be delivered to the Mediterranean instead of ammunition. In the previous chapter, the feasibility of war has been already discussed. But that only means a theoretical capacity to win. Even if war is determined to be feasible, it is not guaranteed that one will actually win. Whether one has 50% chance of victory or 90%, war is technically feasible. Yet, for the relative attractiveness of the option this makes a significant difference. Thus, one has to introduce risk as an additional component, providing a further negative factorto outweigh the benefits of the option.

The key issue in assessing the risks of war is the difficulty of predicting the outcome of battles or wars. This issue has been partially discussed during the feasibility analysis of war. Reliably predicting the outcome of battles is a daunting task due to the inherent complexity of the endeavour. Predicting the outcome of wars is even more difficult. Even if only conceived as a series of battles, the complexity increases exponentially. And battlefield performance alone is often insufficient to determine the outcome of a war. The United States gradually increased its performance against North Vietnam, yet a shift in public opinion resulted in a defeat. To incorporate risk into the analytical framework, this is one of the times when simplification is required: The predictive power of the model is sacrificed in order to achieve utility through reducing the complexity of reality. A proposed way to assess the risks of war is a capacity-based approach. The focus here is not to determine the outcome of individual battles or the probability of something going wrong. A capacity-based approach assumes that something will inevitably go wrong. The question is how many such acts of bad luck one can sustain before friction mounts to the point where it threatens the success of the entire enterprise. As so many things in a realist international system, this comes down to the balance of power. The more favourable the balance of power is to a combatant, the more acts of chance it is expected to be overcome. In contrast, the more equal the balance of power is, the more devastating the influence of chance can be as victory was more in question to begin with. (How to assess the balance of power was discussed in Chapter 12.) For example, a ship breaks down and cannot take part in combat. If one has ten other ships relative to the opponent's two, then this is not a significant setback. If one had only three to begin with relative to the opponent's two, then this threatens the success of the enterprise. Naturally, this is a highly simplified approach and ignores a lot of the minutiae of battlefield strategy. But the future is mostly unknowable; therefore, from a normative perspective such analysis suffices. If the success of a war hinges on perfect execution, then it is a highly risky proposition to begin with and it is probably best to consider alternate foreign policy options. Another thing to note is that capacity-based risk analysis cannot account for black-swan events (high impact/low probability). For example, the scattering of the Spanish fleet near Britain in 1588 is beyond the scope of the capacity-based model as the weather played a significant but unpredictable role in the defeat of the Spanish.

Another risk to war is more related to one's own plans for the war. In most instances, one would want a war to be a limited engagement. This is based on the rational principles of Clausewitz: The effort has to be commensurate to the value of the objective.[6] But there is no guarantee that the conflict will remain within such rational limits. Once war breaks out, controlling it becomes a difficult task. And if it breaks free of one's own rational calculation, one might be forced to go beyond rational costs or even accept defeat. One factor that can affect this is miscalculating the value of an objective for the opponent. Cost–benefit calculations are not only done for one's own side. But one needs to calculate the resistance of the opponent. If the calculation is done incorrectly due to a misunderstanding of the opponent, then it is possible that one expects the enemy to accept defeat at a certain point, but it does not occur. Since they value the objective more than anticipated, they continue fighting, which increases the costs of the conflict. The opponent is not obligated to back down once we assessed that the rational limits of the conflict have been exhausted. Since the international system is governed by imperfect information, this is a constant risk that can only be mitigated through a careful study of the opponent.

Related to this is Gray's argument that in war passion always threatens to overrule rationality.[7] While from an analytical perspective it is useful to consider the international system to be rational, it is not fully so. As discussed before, rationality is a simplification for modelling purposes. It is possible that emotions affect the opponent's strategic thinking and it refuses to cease the conflict. This again can escalate the costs of the conflict, potentially threatening a rational justification for it. History has romanticized plenty of last stands and break-out attempts, primarily due to their heroic futility. Sometimes resistance continues because of the perception that the fate of the entire nation is at stake. Sometimes it continues because of cultural notions of heroism or the shame associated with defeat. Sometimes it continues because of the fear of the consequences of surrender. And sometimes it continues due to a stubborn conviction to waste as many enemy lives as possible. In any case, again, the enemy is under no obligation to surrender when one reached the endpoint of their plans.

And continued resistance can be a major issue. Instead of a quick, sharp conflict to achieve a limited objective, one can find themselves in a protracted conflict where the opponent gradually builds up the means to go on the counter-offensive, eventually threatening one's own home country. Germany expected a quick victory through *blitzkrieg* against the Soviets. Instead, the Soviets managed to turn the tide of the war through a protracted engagement and ultimately reached and occupied Berlin. The termination of a war within rational limits is just as important as accomplishing the objective within the same rational limits. This tendency to escalate is why Clausewitz has been sceptical of the feasibility of limited wars.[8]

When considering the risks, Morgenthau and realism argue that it is best to be risk-averse and adopt a worst-case scenario thinking.[9] For pragmatic risk assessment worst-case scenario thinking, especially to the extreme, can be counterproductive. However, it is equally problematic to engage in wishful thinking and being overly optimistic. Mearsheimer offers a good starting point for assessing a country's appetite for risk: Fear.[10] In offensive structural realism, "fear" is largely defined in the context of the balance of power. The more unfavourable the balance of power, the more insecure or fearful a state is expected to be; hence, it is more willing to accept risks to escape its insecurity. But fear can also contain a significant ideational component. Some states are "scared" beyond what the balance of power itself justifies due to their identity. Insecurity is a cornerstone of China's identity. Its pessoptimist national identity and focus on external threats in its patriotic narrative predispose China to assess its own insecurity to be higher than pure balance of power or geopolitical factors could support. China's assertive, and often risky, behaviour on the South China Sea is motivated by this deep sense of insecurity, combined with a belief in the security of power afforded by possessing significant military might. From a normative perspective, this can be applied in two ways. From the perspective of one's own strategic planning, the more insecure they are, the more it might be worth to take risks to escape it, putting a premium on the attractiveness of such options. From the opponent's perspective, it can help to assess how much risk an opponent is willing to take, which can affect the feasibility and costs of options.

Beyond the theoretical considerations, as always, reputation can offer an important insight. If a country has a history of risky behaviour, then it is reasonable to assume that such behaviour will continue. In contrast, if a country has a history of being risk-averse, such as contemporary Japan when it comes to the deployment of its armed forces, then one can assume that they are unlikely to develop a sudden appetite for major risks.

Understanding the benefits, costs, and risk of use of force is important to establish the overall attractiveness of the option in a given context. However, this is only meaningful relative to other options. Just because an option is attractive, it is not necessarily the optimal course of action. Similarly, an unattractive option cannot be discarded before understanding whether there is anything better. The next chapter looks at alternate foreign policy choices to war and how to make a rational choice about going to war.

Notes

1 Sawyer, R. D. (2007) *The Seven Military Classics of China*. Basic Books: New York.
2 Walzer, M. (1977) *Just and Unjust Wars*. Basic Books: New York.
3 Waltz, K. (2010) *Theory of International Politics.* Waveland Press: Long Grove.
4 Gray, C. S. (2009) *Fighting Talk: Forty Maxims on War, Peace and Strategy*. Potomac Books: Dulles.
5 Levinson, B. (Director) (1997) *Wag the Dog*. Baltimore Pictures and TriBeCa Productions.

6 Gray, C. S. (2009) *Fighting Talk: Forty Maxims on War, Peace and Strategy*. Potomac Books: Dulles.
7 Ibid.
8 Clausewitz, C. (1874) *On War* (Translation by J. J. Graham). Guttenberg.
9 Morgenthau, H. (2005) *Politics among Nations: The Struggle for Power and Peace*. McGraw-Hill: London.
10 Mearsheimer, J. J. (2014) *Tragedy of Great Power Politics.* Norton: New York.

14 Foreign policy alternatives to war and making the choice to go to war

Every problem can have multiple solutions and it is best to avoid the trap of Maslow's hammer. Going to war is simply one tool in the state's foreign policy arsenal to be called on if it is determined to be the optimal course of action. However, for such a determination to be made, it is important to critically examine the various competing options, based on the same cost–benefit–risk model that was used for war in the chapter (Chapter 13) above. This chapter reviews alternative policy courses, including both military and non-military policy options. The chapter also examines delaying, or the conscious choice strategically not to take action. Critical knowledge about the entire toolbox is needed if one wants an effective solution to any problem. The chapter concludes with a discussion of how one can make a rational choice based on all of these considerations.

Military coercion

The primary military alternative to war as discussed so far – which can be termed as the direct use of force due to its focus on directly seizing and securing objectives – is coercion or "diplomacy of violence" as Schelling phrases it.[1] While in execution coercion can look similar to a limited war, it is distinct due to its inner workings. In a traditional war the goal is to seize and secure an objective through military might, and the destruction caused is incidental to it. In contrast, in coercion destruction, and the potential for destruction, is central to the option because the goal is not to directly seize the objective but to persuade the opponent to surrender it in order to avoid damage. It is a negotiation tactic targeting the enemy's will. Using the analogy of the carrot and stick, while bilateral negotiations primarily focus on the carrot of benefits gained from settlement, coercion emphasizes the stick, the damage suffered in case a demand is not accommodated.

The primary benefit of coercion relative to direct use of force is that it can deliver similar *de facto* benefits to war with much lower costs. While it falls short of war, it can mainly deliver *de facto* benefits due to the United Nations (UN) prohibition on threat of force as specified in Article 2.4. However, if it is applied subtly enough and without the need for demonstrative action, then it can potentially deliver even *de jure* outcomes. The costs of coercion are significantly lower because one is not planning a full-scale war against an opponent, only threatens one. Essentially, the opponent is told that if demands are not met a war is unavoidable and that it would cause significant damage to them, far more than what the objective is worth; therefore, it is best if they just give in now. If successful, the objective is secured at far lower costs than what it would have cost to fight the actual war.

The main costs of coercion come from persuading an opponent that a threat is credible, which has two main components: Ability to carry out the threat and willingness to carry out the threat. First, one has to be able to actually inflict the threatened damage on the opponent. As Schelling argues, this often requires one to first overcome the opponent's ability to militarily resist.[2] If the opponent possesses sufficient military might to effectively resist and thus mitigate the damage, then the coercion will not be credible. Visiting the damage of the firebombing of Tokyo is a terrifying prospect useful for coercion. But if one does not possess the means to break through the anti-air defences of the opponent, then the threat is not credible. Second, even if one possesses the means to hurt an enemy, it might be necessary to demonstrate that one is willing to do so. This is the domain of demonstrative actions, short, sharp displays of violence to demonstrate resolve. For example, in a maritime territorial dispute, a demonstrative action can be a short exchange of fire between ships. Importantly, a demonstrative action does not seek to accomplish the overall objective. It simply seeks to demonstrate resolve to engage in violence if the demands are not met. This can make them operationally more flexible than trying to secure an objective through direct use of force.

Building a reputation for resolve is useful for increasing the credibility of threats. But over time a disproportionate reliance on threats has its own reputational considerations. An over-reliance on threats can decrease their credibility. Coercion is best used strategically and sparingly. A nation that relies continuously on threats is often not taken too serious after a time, especially if (a) threats are not actually carried out and (b) the threats themselves appear exaggerated. Countries such as North Korea frequently threaten apocalyptic retaliation, but it is hardly taken seriously beyond preparing for the eventuality that limited demonstrative action or a failed weapons test will take place.

The main drawback of coercion relative to war is the aforementioned need to target the enemy's will. This limits the utility of coercion. Coercion can provide an alternative in scenarios when other forms of persuasion are proved to be ineffective. But it cannot function if the enemy is resistant to the demands. It is possible that one is asking for something that the enemy is altogether unwilling to give or for something that it values more than they fear the threat. In this case coercion will not achieve its objective. While there are many scenarios when the possibility of coercion and war overlaps, coercion cannot fill the genuine niche of war, the ability to resolve a matter unilaterally.

Non-military coercion

The logic of Schelling's diplomacy of violence can also be achieved through non-military instruments. Damage can not only be achieved through the use of missiles and tanks (military power). For example, Chapter 6 has discussed geo-economic statecraft as a potential coercive tool. While in this instance the state relies on economic power, the logic of coercion remains the same as discussed above. To illustrate this, one can take the example of China and the "debt trap" situation often associated with its Belt and Road Initiative (BRI). China has provided significant loans to underdeveloped states such as Cambodia or Laos. China can utilize the leverage in two distinct ways. On the one hand, it can use it to directly acquire assets in the country. China used Pakistan's inability to effectively finance its BRI loans to secure a long-term lease on the port of Gwadar, allowing it to establish a naval base in Pakistan. On the other hand, it can use it to put political pressure on these nations. The Association of Southeast Asian Nations (ASEAN) has been unable to take strong action on the South China Sea due to its consensus-based nature and the reluctance of Cambodia or Laos to take any move that could antagonize China, which in turn could lead to a potential showdown over their large amount of external debt towards China. Essentially, through the coercive influence of indebtedness, China can indirectly interfere with ASEAN even in settings where it is not invited to participate directly. The source of coercion here is the state's ability to cause economic harm. And in an interconnected global economy this can be easily done even without hard power instruments. Another good example would be the 2010 fishing captain incident where China used the threat of a potential rare earth mineral export ban on Japan, whose technology industry depended on China to supply those, in order to secure the release of the fishing captain arrested for ramming a Japanese Coast Guard vessel off the shore of the Senkaku/Diaoyu Islands. Popular boycotts are also common in China, where a domestic audience tries to exert pressure on Japan, or simply expresses its displeasure, through inflicting economic damage by not buying Japanese goods or actively going out of their way to cause damage to Japanese-manufactured cars and Japanese businesses.

Sanctioning a state works very much in the same vein. Targeted sanctions against specific economic sectors seek to hurt the economy of the state beyond other sanctions that might seek to prevent the import of specific goods (e.g., weapons) or technologies. Sanctions are especially effective against authoritarian regimes. De Mesquita and Smith propose that all societies' domestic politics can be described by the relationship between three groups. The *nominal selectorate* refers to all people who have a legal say in the selection of a leader. The *real selectorate* refers to the actual people who determine the leader. And the *winning coalition* refers to the minimum number of real selectorate one requires to win or whose support is essential for winning.[3] In an authoritarian regime, the winning coalition tends to be fairly small and thus it makes political sense on delivering private rewards to them to buy their support, rather than focusing on public goods and good governance to please the masses. Sanctions often seek to directly target the cash flow of specific leaders rather than the country at large. They seek to freeze their assets and sanction-specific sectors of the economy from which they can extract money. This is done in order to cut off the liquidity they would use to pay off the winning coalition, counting on that in the absence of delivering private rewards the coalition will desert the leader, leading to his or her downfall. Sanctions pose economic damage or the threat of such damage to persuade the opponent to comply. In the case of authoritarian leaders, the threat can be a matter of life and death.

With the advancement of technologies, cyberspace has also opened up to coercive politics and the diplomacy of violence. But the logic once again remains the same. Cyberspace simply offers a new method of deliver for damage to act as the coercive influence. Both the civilian economy and military command depend on cyberspace to effectively conduct their businesses. Crippling the stock market, cutting the power grid, severing the link between commanders and troops on the ground, or scrambling the signal between drones and drone operators all present significant problems for a modern industrialized state seeking to conduct war on its own terms. The end result is measurable damage, even if delivered through non-military means. And the threat of such capabilities can be sufficient to persuade an opponent not to risk a confrontation. China has invested heavily into cyber warfare capabilities as trump card weapons to counter US military power.

The costs, benefits, and risks of these avenues are similar to that of military-based coercion. The main benefit is exerting coercion without having to rely on military instruments. The opponent is "persuaded" to comply, even if it is done under duress. In many instances these options can also offer a degree of deniability. Cyberattacks are notoriously hard to trace, offering states a degree of protection from blowback or retaliatory measures.

The main cost of coercion comes from (a) growing threat perceptions and (b) potential retaliatory action. While coercion can be an effective tool for states, it infers a significant reputational penalty. Nobody likes to deal with a bully. Relying on coercion will lead others to identify one as a threat, reducing their willingness to cooperate or offer the benefit of doubt. China's BRI is a good example of this. Because China in certain countries have capitalized on it to gain political influence, the entire enterprise has become tainted and is often viewed as nothing more than an attempt at geo-economic statecraft through the deliberate creation of debt trap scenarios. Not all of China's intentions are malicious and BRI has the potential to deliver genuine benefits to the region. But China's reputation for coercion has made other states wary of participation. Attempts at coercion can also inflict retaliatory damage. One can see this during the US–China trade war. The introduction of US tariffs was followed by retaliatory trade barriers in China. The balance of power is an extremely important measure here: One needs to be sure that the opponent will not be able to retaliate with costs that would be more than what one is willing to pay for coercion. For example, economic coercion has to take into account the balance of power within economic interdependence (discussed in Chapter 5) to ensure that potential counter sanctions will not be more than what one sought to accomplish in the first place.

Just as with military coercion, non-military avenues are situational. If the opponent is unwilling to bend to coercion, then they can be rendered ineffective since they rely on overcoming the opponent's will rather than their strength. A good example is North Korea. The isolated regime remains defiant relying on sanction busting, smuggling, and the regime's willingness to subject the population to misery. Despite the long-term pressure applied on the country, the regime continues to forge ahead with its plans to acquire nuclear weapons.

Peaceful settlement of disputes

Moving beyond coercive power, the international community has developed various options for the peaceful development of international disputes. The primary benefit of pursuing a peaceful resolution option to a dispute is that it allows for the achievement of *de jure* results. Since the dispute is resolved in accordance with international norms, the outcome is generally accepted by the international community. This is a more secure outcome to a dispute and it does not require the continued maintenance by force that characterizes *de facto* outcomes. Acting as a responsible member of the international community also eliminates the reputational penalty associated with military options. Overall, peaceful dispute resolution options can offer significant benefits at relatively low costs. They also do not require the general ugliness and destruction of an armed confrontation. It is hardly surprising that they are preferred resolution methods in contemporary International Relations.

Peaceful dispute resolution options can be divided into two main categories: Binding and non-binding. In a binding option states agree at the beginning of the process that they consider the outcome, whatever it might be, legally binding upon themselves. Whatever decision is reached becomes the new legal reality and disliking it is no basis for not abiding by it – at least in theory. In contrast, in a non-binding process, parties are under no obligation to accept an outcome that they do not find acceptable. Non-binding processes put the emphasis on mutually acceptable compromises whose enforcement is ensured by both parties willingly accepting it. The outcome can be formalized into a legally binding international treaty, but this is not a requirement. These two categories can further be broken down into various procedural avenues.

Non-binding processes can take many forms. The simplest form is bilateral negotiations. Bilateral negotiations essentially involve two states where they try to work out a dispute among themselves through discussion, hoping that they can find a mutually acceptable compromise. Anyone who has tried to negotiate with his/her spouse can appreciate the difficulty of finding a compromise, especially if the dispute becomes entrenched. In a deadlocked dispute it sometimes becomes necessary for an external third party to move along the proceedings. Non-binding third-party interventions can take many forms. In a *good office* a third party acts in a facilitating role to keep the two parties engaged. This can be especially important if diplomatic relations have been broken off. When acting as a *commission of inquiry*, a third party is responsible for establishing the facts of a situation in order to create a commonly accepted basis for the negotiations, but it is not involved in the actual discussions. In *mediation* the third party takes an active role in the negotiation process, but it refrains from recommending a solution to the dispute. The third party takes the most active role in *conciliation*, where it not only undertakes the role of a mediator but also attempts to recommend mutually acceptable outcomes. One can see that the main point of third-party interventions is to break deadlocks and to keep the parties engaged and negotiating. This can be especially important as passions and stubbornness can easily wreck bilateral processes. A key consideration for third-party interventions is to find a person or entity that is acceptable to both parties. They have to command sufficient influence and have to be accepted as impartial by both. The UN Secretary General is a popular choice.

In contrast, binding processes are more formalized. *Adjudication* is most similar to court proceedings in a domestic setting. A decision is made by a deliberating body based on prevailing international law. Just as in court cases the decision is binding on both parties. *Arbitration*, in contrast, is less legalistic and allows for a wider range of factors to be considered. But its outcome is equally binding. Good examples of venues for binding processes are the International Court of Justice (ICJ) or the Permanent Court of Arbitration (PCA). It is important to note that, although the results are binding, states have to consent to the proceedings. If a country refuses to partake in the process, the results are not considered to be binding. The Philippines asked the PCA to establish facts surrounding its dispute with China on the South China Sea. China argued that it does not recognize the authority of the PCA. Thus, while a ruling has been delivered, it is largely symbolic and did not have a significant effect on the dispute. If a state is powerful enough, it can even defy the rulings after the fact. Once the ICJ decided against the United States in the *United States versus Nicaragua case*, Washington argued that the ICJ has no jurisdiction over the United States despite the fact that the United States has been instrumental in setting up the ICJ as part of the broader UN framework after World War II. While the United States eventually agreed to observe the ruling, it took decades and the decision was only made because constant reminders from the General Assembly were becoming more bother than the case was worth.

In deciding which avenue to utilize, it is important to weigh their relative strengths and weaknesses. A binding process will create a clear and binding conclusion to the dispute. However, one needs to give up control over the process. If one's claims are not strong enough, there is a good chance that one will walk away with less than the desired outcome or even with none of the objectives accomplished. Since the binding process puts a conclusive end to the dispute, there isn't much room to manoeuvre once an unfavourable ruling has been delivered. In contrast, non-binding processes offer more control to the participants. However, with more control come more pitfalls. It can be supremely difficult to arrange for a mutually acceptable compromise. The process is prone to deadlocks and breakdowns. Non-binding options often require a very long time to successfully conclude and can be dependent on successful confidence building between the two parties. And even once the process is concluded, due to the need for a compromise, one is expected to walk away with less than what one was seeking to accomplish. A good example of this is joint development. While one gets access to some resources, they need to be shared with the other country. Better than nothing but less than if one established control over the disputed area. Non-binding processes also allow for more interference and coercion. In a binding process impartial arbiters deliver a decision. But in a non-binding process it is possible to exert pressure on the participants through various means of diplomacy, including the coercing influence of geo-economic statecraft. It is not an accident that China has a strict insistence on bilateral negotiations where it can maximize its relative power advantages and can divide countries through the strategic use of concessions in a multi-party dispute.

From a reputation perspective, peaceful conflict resolution options offer both benefits and constraints. Resolving disputes peacefully increases one's reputation as a responsible member of the international community and lowers threat perceptions. This can make it easier to engage in cooperation with other states. To put it into constructivist terms, a pattern of peaceful conflict resolution can shift a country's relationship in the cooperative/neutral direction. However, engaging in peaceful conflict resolution also sets the expectation that one will do so in the future. This has a few implications. The international community will expect similar methods to be used in the case of other disputes one might not want to resolve peacefully due to weaker claims or the costs of compromise, including potential domestic political backlash. This is especially so for binding peaceful resolution options. Once one has used, for example, the ICJ, it is hard to justify why one would not recognize the jurisdiction of the court in another case. While using these avenues can offer reputational benefits, defying resulting expectations that peaceful conflict resolution will be a larger pattern of behaviour can impose reputational penalties. If one seeks to build a diplomatic strategy around a favourable reputation, and its ability to enable cooperation, then these reputational penalties can be a significant drawback. Essentially, from a reputation perspective, if one seeks to engage in peaceful conflict resolution, it is best to commit to it as a default, rather than on a case-by-case basis.

Relative to use of force, the main drawback of these peaceful resolution options is similar to that of coercion: One cannot ignore the opponent. In case of non-binding options, one needs to reach a mutually acceptable compromise. But even in the case of binding options one needs to secure at least their consent to submit the case to a deliberative body. If one is facing a stubborn opponent that is unwilling to engage in these avenues, such as China that has shown itself to be on occasion, then the utility of these avenues is diminished. Just as coercion, peaceful resolution options cannot fill the small but significant role of use of force in international relations. In the day-to-day running of the international system, they can offer a cost-effective solution to many international disputes but are often inefficient dealing with black-swan international conflicts, where the stakes are high but negotiations proved to be ineffective. War remains the only option where one can accomplish something largely unilaterally, assuming that one possesses sufficient power.

Delaying

One final alternative course of action is the course of inaction. Delaying refers to the strategic choice not to take action. The outbreak of World War I has partially been credited to the "cult of the offensive" that was present in contemporary military doctrine. Military thinking then has assumed that action is always preferable to inaction to maintain the initiative. This thinking has been criticised by Corbett in his defence of a defensive naval strategy.[4] Sometimes the optimal course of action is simply to do nothing. To be technical not doing nothing, but to delay making a decision. This can be illustrated with a simple example. China seeks control over the Senkaku/Diaoyu Islands. Peaceful dispute settlement options are not attractive: In case of arbitration or adjudication China would only have a rather weak legal case and Japan denies the existence of a dispute, so Tokyo would be unlikely to engage in bilateral negotiations. Japan has also proved itself increasingly resilient to various forms of coercion as Tokyo grows visibly tired of China's assertive conduct and has shifted Japan towards a more deterrence-focused posture under Shinzo Abe to directly counter China. This leaves the use of force: If Beijing desires the islands it would need to go to war, a limited one but a war nevertheless, to seize the islands. However, the balance of power at the moment is not favourable for such a course of action. The Japanese Maritime Self-Defence Forces remain a superior naval force. Under such conditions the smart policy approach is not to forge ahead with war nevertheless, believing that inaction is not an option, but to delay.

Delaying is not inactivity, simply inaction in a specific decision-making context. China could choose to defer a resolution to the dispute, utilizing the time to try to affect the situation. China could use the time to build up its navy to achieve a more favourable balance of power with Japan, increasing the chances of success in a military confrontation. Beijing is seemingly pursuing this route, albeit not primarily motivated by the Senkaku/Diaoyu Islands dispute. Alternatively, China could seek to remove obstacles from peaceful resolution options by adopting a more conciliatory tone towards Japan in an attempt to redefine the relationship. In any scenario, securing more time could be beneficial in increasing the feasibility of specific options, creating a future position where a resolution is easier.

The main cost of delaying is one of perception. Most countries invest a lot of political capital into their disputes, especially when it comes to domestic mobilization or to take advantage of the "rally around the flag" effect for domestic political gains. While delaying can be a strategic choice, with activity often hidden behind state secrecy, it can look like inaction from the outside. It can trigger a domestic political backlash if the domestic audience becomes impatient or becomes to view the government as weak for its lack of response. China has sometimes experienced this in the context of the Senkaku/Diaoyu Islands dispute. The Party has invested a lot of political capital into its nationalist narrative, including the recovery of lost territories. However, sometimes it is beneficial to adopt a delaying approach. This can inflame tensions in China where events related to China's disputes, especially those involving Japan, can trigger mass demonstrations and civil disorder. Such displays of domestic anger put pressure on the Party to pursue a hard-line approach as their primary concern is domestic stability and regime survival. A similar effect can happen on the international level. In various chapters this book has discussed the importance of reputation for international relations. By frequently adopting delaying, especially if it is accompanied by sabre rattling, states can acquire a reputation for weakness: That their threats are empty and the government is not actually ready to follow through. This can lower the effectiveness of various foreign policy options, most importantly various avenues of coercion.

While delaying is usually a low-cost, low-reward option, it does come with significant risks. Delaying is pursued with the understanding that the decision-making environment will change in positive ways over time. But there is no guarantee for that. It is possible that the future position of the state will be even weaker. For example, a state can choose to delay as the balance of power is unfavourable. The hope is that it will be able to undertake a military build-up to favourably affect the balance. But there is all the possibility that the opponent will not only keep pace but potentially outperform one in the ensuing arms race. Furthermore, the future is inherently uncertain. There is no guarantee that "black swan" events will not crop up. A sudden political crisis can undermine the government's legitimacy, reducing domestic stability and the country's ability to focus on foreign affairs. Events such as a sudden economic crisis or global pandemic can destroy delaying-focused policy avenues.

Overall, it is important to not fall victim to a cult of the offensive, choosing to act when all options are significantly unfavourable just to see a result. At the same time, it is crucial to recognize the fragility of the future and that one cannot trust that the benefits of delaying will be delivered. Sometimes it is better to act from a position of weakness than wait for the situation to further deteriorate. The question of delaying raises the importance of normative research and forward-looking policy formulating, which incorporates the future as a domain of consideration.

The rational decision to go to war

The final step in the rational choice model is to make the actual decision, in this case whether to go to war or not. This chapter focuses on how that choice can be made, concentrating on the comparison and contrast of competing policy options based on their cost–benefit–risk ratio. The section then proceeds to examine how this choice is altered if one moves from the perspective of a single choice to the perspective of a multi-level game. Foreign policy formulation rarely focuses on singular choices. Rather it is better characterized as a reciprocal multi-level game. The rational choice model described so far can facilitate the making of a single decision. But for it to be an effective tool for foreign policy formulation it has to be contextualized within this larger multi-level game, especially as long-term considerations might affect the attractiveness of options, altering the rational choice ranking.

Making the choice

In the above sections, we have established a potential feasibility set for a foreign policy problem at hand. For the rational choice model, we analysed four main potential foreign policy options: Use of force, coercion, binding peaceful resolution options, and non-binding peaceful resolution options. The question then is which one is the rational course of action. To determine this one needs to compare their relative attractiveness based on their benefit–cost–risk ratio. In a generic format, these can be summed up as shown in Table 14.1.

Table 14.1 Option comparison

	Use of Force	*Coercion*	*Non-Binding Peaceful*	*Binding Peaceful*
Benefit	*De facto* results can ignore enemy's will Reputation for resolve	*De facto* results Potential for *de jure* results	*De jure* results reputation for responsible member	*De jure* results reputation for responsible member
Cost	Material costs of war International response Reputational penalty	Cannot ignore enemy's will Cost of demonstrative action Growing threat perception	Need for compromise: Less than 100% accomplished Potential domestic backlash Expectation for future behaviour	Lack of control over proceedings Expectations for future behaviour
Risk	Military defeat Failure to end conflict within rational limits	Credibility Escalation, potential armed confrontation	Deadlocks and breakdowns	Decision made against Less than desired awarded

Table 14.2 Symbolic coding

	Benefit	*Cost*	*Risk*
Low	+1	–1	–1
Moderate	+2	–2	–2
High	+3	–3	–3

Table 14.3 Attractiveness of war

Benefits	
De facto Control (Moderate)	+2
Can Ignore Opponent's Will (High)	+3
Reputation for Resolve (Low)	+1
Costs	
Material Costs (Moderate)	–2
International Response (Low)	–1
Reputational Penalty (Low)	–1
Risks	
Military Defeat (Low)	–1
Failure to End Conflict (Low)	–1
Total	–1

One semi-quantitative way to formalize these calculations is to attach a numerical indicator that sums up the qualitative arguments of the research. A potential way to do that is to create set categories as shown in Table 14.2

The categories are coded with a numerical value that represents the assessment of the observer. Then the balance can be simply expressed by the sum of the numerical representation of each option. An example of this is shown in Table 14.3.

Based on Table 14.3 one can see that the attractiveness of war – quantified from a qualitative argumentation – is not particularly high. But this has to be compared with the attractiveness of other options. Ultimately it is not total attractiveness but relative attractiveness what matters. The international arena is full of decision-making situations where one is forced to choose between largely unfavourable policy options. Even in those situations a choice nevertheless has to be made. A rational choice approach can help one to identify the most optimal course of action.

The quantification of the variables might appear highly arbitrary. And to some degrees it is. However, it is important to remember that one is creating a numerical representation of a complex reality. Without the qualitative reasoning behind it, the numbers, and the analysis itself, are meaningless. To use the language of quantitative statistics, the goal here is to create a function

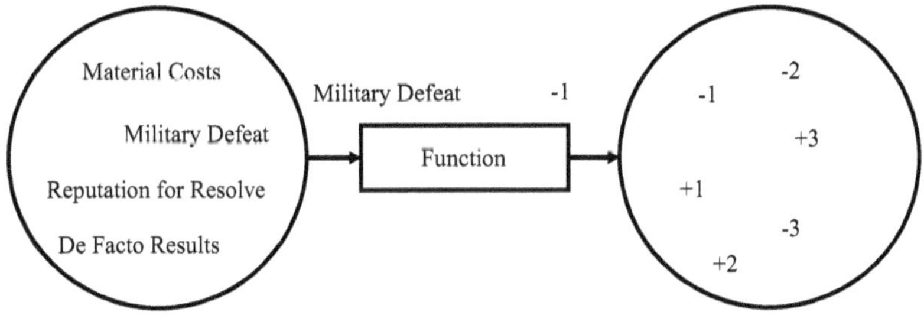

Figure 14.1 Function

that can take a complex qualitative argument and translate it into a numerical value. A function is a rule or formula that can translate elements of one set into elements of another set. In this case, the first set is comprised of qualitative conclusions one has reached about the various factors that affect the attractiveness of the option, while the other set consists of numerical data. The function ensures the translation from the first set to the second set.

This function can be complex or simple. A simple function is sufficient if the goal is only to create a visual representation for the argument, such as creating charts and graphs, but the analytical focus remains qualitative. If one wishes to engage in more demanding quantitative analysis afterwards, the function has to be able to capture the nuance of the qualitative arguments sufficiently and thus more critical engagement is needed with the function itself. Either way, at the end of the rational choice process one should possess a clear idea on the optimal course of action, from a purely rational perspective.

While one now possesses a purely rational answer, there are some caveats one has to observe before implementing the policy option in real life. First, as discussed earlier, it is possible that an option has to accomplish a specific objective, otherwise it cannot be chosen. This can be largely eliminated during the feasibility testing. But if one did not control for this during the feasibility testing, then it is important to critically examine the emerging optimal course of action and whether it violates some existing rule or requirement one has for the acceptable policy course. Second, Mearsheimer's offensive structural realism offers an interesting consideration. Mearsheimer recognizes the need for rational policy formulation. But when it comes to selecting the optimal course of action, he argues that one has to take into account how others will benefit from the decision. Offensive structural realism argues that even if an option can deliver substantial benefits to one, if it also delivers more benefits to somebody else, then one should not pursue it as the overall affect in the balance of power will still be a negative shift.[5] This raises the question of playing a multi-level game and applying game theory to foreign policy formulation, which is going to be the focus of the next section. But it also raises the idea that foreign policy decisions don't take place in a vacuum.

The international system is interlinked. Decisions in one situation will have a ripple effect throughout the system. For effective policy formulation these have to be anticipated and assessed as well, adding an additional factor to the attractiveness of a decision. Ultimately, such considerations can be inserted into the numerical model as an additional cost or benefit factor – based on whether it benefits or inconveniences another state – as discussed above.

If one looks at war in generic terms, one is likely to find that the use of force is a niche foreign policy option. It has inherently high costs, meaning that one would be prudent not to be too quick to resort to use of force. War should be always viewed as a last-resort option. However, under certain circumstances, primarily when all other options are inhibited by various factors, it can offer a way to resolve a policy situation. Thus, even if viewed as a last resort, it needs to be considered critically in a policy situation.

Playing a multi-level game

So far, this chapter has focused on making a single rational choice. However, this is not necessarily representative of the reality of international relations. This has been hinted at through the repeated mention of the role of reputation. If international relations are understood as a series of single decisions, then it is an oversimplification of the model that can create inaccurate policy assumptions. A single-level game is only concerned about the costs and benefits of that decision. Utility maximization needs to be achieved through single steps. For example, if one takes the Senkaku/Diaoyu Islands dispute, and if it is conceived as a single-level game, then the use of force makes sense as other policy options are blocked by circumstances and it would allow the establishment of *de facto* sovereignty that is sufficient to fulfil China's security interests. However, this assumes that this decision is made largely in a vacuum. The decision as well as the costs and benefits are largely immediate. However, this is not how international relations realistically work.

In contrast, one can envision international relations as a long-term multi-level game: A chain of decisions where each player reacts to the behaviour of the other. If conceived in such a manner, then it will have significant implications for strategy and policy decisions. The way to "win" obviously changes if one has to achieve an objective in a single move versus if the game is played over multiple rounds. One easy example is to conceive in the context of the prisoner's dilemma: One is arrested with another person, both taken to separate interrogation. If one betrays the other, then they receive no punishment but the other does 10 years. If neither betrays the other, then both serve 5 years. If the decision is envisioned as a singular event, then one can maximize utility by betraying the other. However, if the decision is envisioned in a larger context, then the other person will have to make a choice as well, in which case he/she also has an incentive to minimize jail time. If both persons betray the other, they both serve prison time, the worst possible outcome from their perspective. Adopting a short-term focus by viewing the game as a single-level contest can produce suboptimal results in the long term. This can be further complicated if we view the interrogation as a series of questions where one has the opportunity to hold out or partially incriminate the other but also reduce suspicion on oneself. Essentially a single-level focus emphasizes zero-sum thinking and competitive behaviour because the advantages of cooperation are usually revealed only through the long term. In contrast, in a long-term multi-level game there is a reduced emphasis on immediate utility maximization and players can afford to take short-term losses in order to secure higher gains in the long term.

To return back to the Senkaku/Diaoyu Islands dispute, as said before, if it is conceived as a single-level game, then there is a push towards a military resolution. However, if it is viewed as a multi-level game, then delaying becomes a viable option. In a single-level structure delaying is not meaningful: It achieves nothing, so the decision does not accomplish any objectives. However, in a multi-level game it becomes a vital tool in international relations. Delaying allows a state to "pass up" its turn, hoping that conditions will evolve to be more favourable over time for a resolution. A long-term perspective dispels the notion that action is always superior to inaction.

It is important to note that international relations are not static. Both the context and the players evolve. This is especially important in a system governed by imperfect information where much of one's actions has to be based on assumptions. Observing another player can provide valuable insights into their behaviour, especially if critical events arise that they need to respond to. Expanding the timeframe also allows one to wait for the overall context of the game to change. To return to the example of prisoners, if one knows that the police can only hold suspects for 24 hours, it makes sense to delay, because in the absence of evidence they would need to release the suspects, allowing for an escape. To borrow a term from probability statistics, conceiving international relations as a multi-level game allows states to engage in Bayesian updating. At its core, Bayesian updating refers to the idea that if new information emerges, then the probability of events changes. The classical example of this is the Monty Hall problem. In the Monty Hall problem, there are three doors, with one having a price behind it. One needs to choose one door, then another door will be opened, and then an option will be given to change one's selected door before the location of the price is revealed. From a probability perspective it makes sense to switch doors after the first one has been opened because new information emerged – by opening a specific door the gameshow host has a probability of doing so because (a) the other non-selected door contains the price or (b) a random choice – that affects probability. To translate this to international relations, in a multi-level game, states can observe the behaviour of each other and learn new information from it. This can affect their decision-making model by updating how their behaviour will affect the accomplishment of objectives over multiple rounds of the game. Essentially, instead of viewing decisions of vacuum, one can conceive international relations

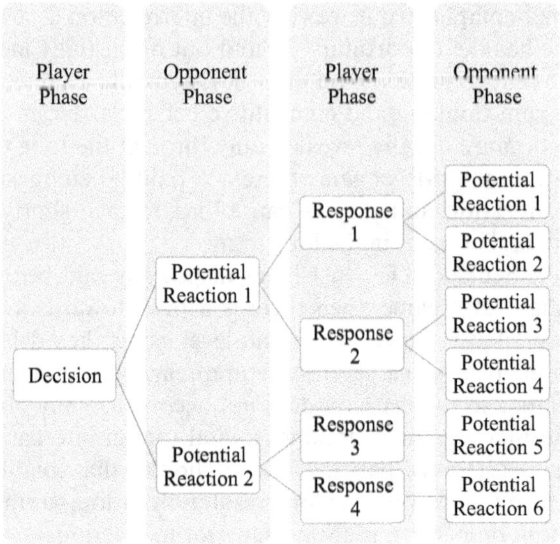

Figure 14.2 Multi-level game

as a chain of reactive decisions, where the players keep responding to each other's actions, as shown in Figure 14.2.

If one looks at Figure 14.2, one can see how such a multi-level game would work. Through a series of rational analyses one can establish predictive analysis on how the opponent is probable to respond and then how it affects one's policy options going forward. The important thing about visualizing the decision-making situation as such is to highlight the importance of long-term planning – because the objective will be at the other end of the decision-making tree, not at the next step. Thus, one can afford, or, in some case will be forced to, accept losses at certain steps, if it nevertheless builds towards the accomplishment of the objective.

Another important reason to visualize the decision-making tree is to appreciate the uncertainty encoded into it. Remembering Gray's maxim that the enemy has a vote too,[6] one can see that at any level of the game the opponent has multiple ways of responding. Rational decision analysis can help with establishing the probable course of their action, but it cannot offer certainty. One needs to not only map out the ideal path of decision-making, but also set up contingency planning for the eventuality that the opponent will deviate from predictions. This brings a key point into sharp focus: The longer the game goes on, the more uncertainty is introduced. The more levels are to the game, the harder it becomes to establish the probability of behaviour and predict an outcome. This acts as a direct counterforce to what has been said earlier. Focusing on a long-term game allows a country to learn the behaviour of the opponent and to pursue goals more effectively by reducing short-term pressures to be competitive. However, the longer the game goes on, the harder it becomes keeping it on track to actually realize the end goal. And if it breaks down, then all the short-term sacrifices were for naught. The nature of probability suggests that the shorter the game, the more predictable it is. Simply looking at it from a mathematical perspective, at level 1 the probability of an outcome (assuming that there is only two possible outcomes) is ½ to ½. By the next level, the probability of an outcome is 1/8. Now this simplistic model ignores issues such as overlap between outcomes or multiple acceptable outcomes, but it is a stark reminder of the effects extending the game can have.

This means that states need to balance the length of the game. The accomplishment of objectives cannot be viewed as a single-level game, but it also cannot be a 125 years plan. Also, states need to be mindful that they don't suffer a death by thousand cuts: If not managed carefully, small costs incurred gradually can nevertheless mount to outweigh the benefits of the overall policy objectives. This is especially true in the realm of security, where short-term acceptance of negative outcomes can spell disaster and embolden opponents, decreasing the probability of whatever long-term outcome one hopes to accomplish. One can illustrate this with a simple hypothetical scenario: Developing a doomsday weapon. If developed, the doomsday weapon would grant ultimate security. However, it takes a long time and a lot of resources to develop. If one accepts it as a long-term goal, then one might need to cut back other military spending and follow appeasement to opponents. While adopting this strategy, one's own capital might be overrun by the enemy before the weapon is completed. This hypothetical scenario can be partially seen in the German *Wunderwaffe* programmes during World War II that focused on the creation of super weapons at the expense of more immediately usable equipment. Reasonably speaking, by the end of the war only a miracle could have granted Germany victory, and no regular equipment would have been enough, which supports the German efforts, but nevertheless it is an illustration of how such rational decision-making concepts can play out in real life.

Overall, adopting a multi-level perspective can improve how realistic one's model is and thus its predictive or explanatory capabilities. However, applying game theory is not without its methodological challenges. While game theory is primarily the domain of probability mathematics and statistics, in an international relations context it involves a high degree of qualitative research and reasoning due to lack of possibility for experimentation. Looking at probabilities and future forward-looking policy planning in the international system, especially if done through a qualitative methodology, involves a high degree of subjectivity. One needs to critically engage with a lot of non-quantifiable factors where reasoning and interpretation are needed. This means that, as an analytical model, the contents of this book are best used in a normative context: They work best in creating future policy. Used in an explanatory context, it maintains the ability to produce valuable insights, but it won't have a 100% effectiveness as reality crashes against the constructed model. Nevertheless, if applied correctly, critical rational decision-making offers a better shot at success than simply trying to go with one's guts or to rigidly follow pre-existing theoretical assumptions. While this might not sound like much, one has to appreciate the complexity of the international system and the importance of forward-looking policy planning under the conditions of anarchy and imperfect information. There is no silver bullet to foreign policy formulation. But to use a metaphor, just because equipment failure cannot be eliminated, it does not mean that one should jump out of an aircraft without checking the parachute first.

One additional consideration, which is becoming a running theme of the book, is reputation. Understanding one's opponent is crucial in determining how it is likely to act at any point in the game. If an opponent has a cooperative reputation, or a cooperative relationship with us specifically, then it increases the probability of cooperative behaviour in the game, unless one adopts a course of action that would actively damage the cooperative relationship. In contrast, if an opponent has a reputation to be competitive, or has a competitive relationship specifically with us, then it is reasonable to assume that they are more probable to adopt competitive responses. But one's opponents will also engage in their own Bayesian updating, which opens the door for a constructivist strategy. If one offers a more cooperative stance, all things being equal, it is reasonable to accept that a rational opponent will shift its behaviour accordingly, creating a foundation for changing the nature of the relationship. However, it is important to be mindful of (a) it being a very long-term possibility and (b) the need to avoid being taken advantage of in the short term. Competition is expensive and if one can achieve an objective cooperatively, then there are limited reasons not to do so. The main obstacle is the realist notion which suggests that if somebody else benefits greatly in the balance of power, then a course of action is not worth it even if one can also benefit from it.[7] The key issue for International Relations is that, in most instances, a cooperative stance requires a difficult compromise.

In the end, adopting a multi-level perspective is necessary to make the decision-making model reflect reality better. But it is important not to get lost in the long-term promise of a multi-level game. In most decision-making situations, a result is needed within a reasonable time. Thus, rather than short- or long-term planning, one should focus on the medium-term planning, while in doing so balancing both short-term and long-term considerations. This is especially so in a democratic country where the four-year election cycle means that, one the one hand, results need to be delivered for political survival, and on the other hand, leadership can change in every four years, reducing predictability. One only has to look at the policy reversals undertaken by the United States after President Trump took office following President Obama to appreciate this.

Notes

1 Schelling, T. (2008) *Arms and Influence*. Yale University Press: New Haven.
2 Ibid.
3 De Mesquita, B. B., and Smith A. (2011) *The Dictators Handbook: Why Bad Behaviour Is Almost Always Good Politics*. Public Affairs: New York.
4 Corbett, J. (1911) *Some Principles of Maritime Strategy*. Guttenberg Online.
5 Mearsheimer, J. J. (2014) *Tragedy of Great Power Politics.* Norton: New York.
6 Gray, C. S. (2009) *Fighting Talk: Forty Maxims on War, Peace and Strategy*. Potomac Books, pp. 134–135.
7 Mearsheimer, J. J. (2014) *Tragedy of Great Power Politics.* Norton: New York.

Part IV

Conclusion

A tragic but necessary evil

War is terrible. There is no escaping that. But one does well to remember the words of Joseph Stalin. Stalin argued that, while the death of a single person is a tragedy, the death of a million is statistics. The quote is often used to illustrate the uncaring, almost nonchalant, brutality of the Soviet regime. But a closer inspection reveals a deeper truth. The death of any individual, understood in the context of their life, is a tragedy. However, the policymaker does not have the luxury to view the world from that perspective. The policymaker views the world from an overhead perspective where the stakes of the game are the fate of nations and millions. It is a dehumanizing perspective in which the uniqueness of the individual that underpins the tragedy of their death is extinguished. This is necessary to allow the policymaker to make the hard decisions where the lives of the few have to be sacrificed for the benefit of the many. This does not diminish the tragedy of the death of the individual. But knowing the stakes, such tragedies are on occasion required. The decision-maker's responsibility is to formulate policies that best serve the nation, even in violation of one's conscience. In a system of life or death decisions, morality is the luxury of the living.

War serves a specific purpose in the foreign policy arsenal of the state. Its purpose is to accomplish objectives that are vital to the nation but proved to be resistant to all other forms of policy. Only then can one justify the high costs in blood and treasure required to engage in war. Rationality dictates that we view war as a tool of last resort. But it also dictates that we nevertheless consider it when formulating foreign policy. This chapter has provided a detailed framework on how to analyse the rational decision leading up to war. It walked through the key phases of the rational choice analysis model and explored how the specific considerations of war can be integrated into such a model. Utilizing the model provides an analytical framework of why a country engages in war in a supposedly rational international system. It also offers tools to explain why under different circumstances it is not rational to resort to the use of force. As a normative analytical framework, it can be useful for both policy recommendations and for anticipating the behaviour of others. Combined with the previous three parts, Part IV provides a comprehensive overview of the occurrence of war in the international system.

Understanding war is essential, even if one considers war to never be an acceptable component of the international system. Effective preventative diplomacy is dependent on the ability to understand the strategic calculations that might lead a state to war and identifying key "attack points" in those calculations where effective policy can strategically raise costs or provide incentives for alternate behaviours. Understanding these strategic calculations offers an opportunity to targeted surgical deterrence, exploiting specific components of the opponent's strategic thinking to prevent the occurrence of war. We would all like to live in a peaceful international system. But peace starts with the understanding of why we go to war. Even if one is not interested in war, war is always interested in them.

15 “A” peace versus “our” peace

With potentially reaching the rational decision to go to war in the last chapter, this book’s examination of war comes to a conclusion. This is not a book of strategy or tactics, and it does not intend to examine how to fight a war or more importantly how to win a war. On that topic, this book simply would offer the advice to only go to war if one is certain that the political objectives one seeks can be realized through war in the first place. The purpose of this book has been to study the occurrence of war in an International Relations context, especially focusing on a policy formulation perspective. Through the last four parts the readers should have a broad understanding of the concept of war, the theories surrounding its continued occurrence, and the rational choice methodology applicable in a policy context. This should offer a solid foundation to begin one’s own exploration of the subject, including case study analysis and normative policy research. This final chapter reflects on some key points made in this book as well as offers a case study of the Senkaku/Diaoyu Islands dispute to illustrate how the rational choice model can be applied to conflict analysis.

War as a tool of statecraft, far from obsolete

One of the key points of this book has been to portray war (use of force) as a tool of statecraft, a part of the state’s foreign policy arsenal, rather than something distinct from politics or diplomacy. Part I of the book focused on describing war as a concept in service of this idea. Chapter 2 specifically highlighted how war is often interpreted as something distinct from politics. This is often based on the misinterpretation of Clausewitz’s quote on how war begins where politics ends. This book argues that war is not distinct from politics, or international relations, but, as Clausewitz put it, the continuation of the same political commerce. War is inherently about politics: It is about realizing political objectives through the medium of military power. Forgetting this usually leads to catastrophic policy failure. An emblematic example to this is the United States (US) invasion of Iraq in 2003. There was a strong disconnect between the military phase and post-conflict phase, resulting in a failure of translating military victories into the accomplishment of political objectives. The moral objection to this idea is fully understandable. War is always terrible. It is a natural inclination to consider it as something separate from the everyday business of international affairs. However, it is important that such moral objections don't obscure the nature of international politics, leading to improper policy formulation. War has been with humanity since the first tribes have formed and will likely stay with humanity until the last remnants of society are extinguished. It does so because it occupies a unique space within the foreign policy arsenal: The use of force is the only key way in which an objective can be pursued without targeting the enemy's will in some form of persuasion. War is the only key way in which a thoroughly uncooperative opponent, who responds neither to persuasion nor to threats, can be subdued. No other tool within the state's vast arsenal can lay claim to this role. Otherwise, war being as terrible as it is, humanity would have forgone its use.

Some do claim that the time for war is over, that it is a relic of a less enlightened age, made obsolete by advancements within the international community. This argument would be akin to claiming that murder has been eliminated from a domestic system. These arguments ignore a simple equitation: Conflict of interests has not been eliminated from the international system and war offers a unique resolution to disputes that often cannot be replicated by other tools, thus declaring the demise of war is premature. Periods and pockets of peace do occur. But they are neither all-encompassing in scope nor permanent in duration. The end of the Cold War has ushered in an unprecedented time of unipolar order, where globalization seemingly brought forward the victory of a single ideological position. All major opponents have been dispatched and the world organized into a single system. However, this did not eliminate wars. Various conflicts have emerged that resulted in the use of force. And over time new challengers have emerged with their own vision for the international system, raising the spectre of major power wars. Part II of this book examined the various theories of International Relations that have argued for the obsolescence of war or could be interpreted as doing so. The results in each instance have been clear: While one can argue that the space for war is limited, there seems to be no reliable way of totally eliminating that space. This connects back to realist arguments on how the system is fundamentally conflictual. As long as there are multiple visions for the rules of the international system and domestic governance, conflict will be unavoidable. This is exacerbated by the inherent scarcity of resources and space that humanity experiences confined to the planet. Some of these conflicts will be resolved peacefully. Some of them could be resolved peacefully, but will not be because of imperfect information, miscalculations, self-interest, or simple stupidity. And some conflict will be inherently resistant to a peaceful resolution due to irreconcilable differences, necessitating a test of mettle to determine the outcome.

If one accepts the premise that war continues to be a part of international politics, this book then proposes a rational choice-based framework for its analysis. Parts III and IV present a critical examination of the reasons one might consider going to war and how such a decision is made in a rational choice context, respectively. A focus on rational choice modelling is important because of its utility for normative research. Devising strong policies is essential to international politics: It is much better to devise strong forward-looking policies than to provide in-depth explanation as to why a policy approach failed in retrospect. Policy formulation is inherently forward looking – its goal is to deal with contingencies that have not materialized yet. Thus, it is a domain for normative research, both in creating ideal behaviour scenarios and attempting the arduous task of anticipating the behaviour of opposing states. This is sometimes dismissed by the more historical minded as "soft science": How could one reasonably attempt to analyse an uncertain future? A rational choice model argues that actors behave in a relatively predictable way, allowing one not only to explain past behaviour but also to discern future intentions. While this approach is not without its limitations, it offers a more solid ground than trying to simply guess behaviour or shunning the enterprise altogether. Part IV is specifically dedicated to exploring how such a behavioural model can be applied to the subject of war.

One of the key points of this book is to present war as an integral part of international politics that is still a relevant field of study, even if one's only goal is to achieve peace. The next section will focus on presenting a practical case study of how preventative diplomacy can be analysed and formulated through this framework. Understanding war in a rational context is going to be critical to devising effective preventative policies through targeting critical points within states' foreign policy calculations. This book would argue that such preventative planning is the most reliable way to minimize the occurrence of war within the international system.

Preventative diplomacy on the East China Sea (case study)

This book so far has looked at various considerations for policy formulation as relating to war in general terms, using various examples to illustrate the key points. However, as a conclusion it would be valuable to see how all of this comes together in a single case scenario. To do so, this concluding chapter offers a skeletal overview of how these rational calculations would work in the context of the Senkaku/Diaoyu Islands dispute between China and Japan.

The Senkaku/Diaoyu Islands dispute is one of the most contentious territorial disputes in Asia. The dispute has seen several cycles of escalation, some of which presented a potential for a military confrontation. The dispute is emblematic of the regional tensions caused by the rise of China as Beijing pushes against the established status quo to carve out its own sphere of influence. The islands themselves are remarkably unremarkable: It's a group of small barren landmasses, rocks and reefs. There are no human inhabitants

Table 15.1 Chapters

Chapter	*Thesis*
2	War is a tool of statecraft within the same political commerce as diplomacy rather than distinct from it. Traditional interpretations focus on the outcome or medium of war, rather than the fundamental nature of it.
3	National identity and culture are important components of war. A nation's histories, and the identities they create, play a key role in how they view their security and their capacity for war. Similarly, nations can possess unique strategic cultures that affect their willingness to engage in war and how they would fight such wars. These have significant implications for the analysis in later chapters.
4	Humanity has often found morally objectionable and has attempted to devise various legal regimes to control or eliminate war from the system. These endeavours have not resulted in the abandonment of war.
5	Theories such as defensive structural realism or economic interdependence theory argue (or are often interpreted to argue) that war is a self-defeating enterprise. While they can offer an argument of why some wars should not occur, they do not eliminate war from the system.
6	Smart power and geo-economic statecraft are often interpreted to offer viable alternatives to war that can accomplish the same at lower costs. While they are valuable tools, they cannot fill the niche occupied by war.
7	Social constructivism and Mueller argue that fundamental evolution in international politics can lead to the realization of a peaceful, non-competitive/non-conflictual international order. These arguments offer questionable pragmatic utility as war continues to persist in the system.
8	The occurrence of war necessitates the permissive presence of structural factors such as an imbalance of power or a desire for global rule setting.
9	National characteristics such as a disconnect between physical and political geography, national identity, and the nature of domestic governance can contribute to the occurrence of war.
10	While various factors mentioned in the previous two chapters are important for creating the conditions that permit the occurrence of war, the actual trigger for a conflict is often something more concrete, such as the escalation of territorial disputes or a potential domestic diversionary motivation.
11	If states are considered to be rational, then their behaviour can be modelled through rational choice analysis. However, such an analysis has to be based on bounded rationality.
12	Determining the feasibility of war requires complex considerations beyond the total amount of military power held by the state.
13	Cost–benefit–risk analysis is at the centre of determining the optimal course of action. There are multifaceted factors that can affect the utility of war.
14	The rational decision to go to war is made based on a comparison between the utility of war and alternate foreign policy options.

or significant vegetation. The most advanced inhabitants are indigenous moles and ants. Despite this the islands are at the centre of fierce competition between China and Japan, the former engaging in an assertive campaign to press its claims over the islands. There are a number of reasons for this:

- **Economic**: The East China Sea is reported to be rich in natural resources and the Senkaku/Diaoyu Islands play a role in the delimitation of exclusive economic zones under the United Nations Convention on the Laws of the Sea (UNCLOS). The distance between the Chinese and Japanese coasts is less than 400 nautical miles (2 × 200 nm or the maximum extent of the exclusive economic zone [EEZ]) and thus China's and Japan's claims overlap. Japan argues that delimitation should be based on an equidistant meridian line, while China claims that the Okinawa Trough represents the natural border of the continental shelf, thus it should be allowed to extend its EEZ until then. The Senkaku/Diaoyu Islands play two key roles in this situation: (a) The islands can potentially claim their own EEZ, affecting the ownership of fish, oil, and natural gas in their surrounding, and (b) whether they are part of Japan's coastline can affect the delimitation in a larger context.
- **Political**: The Chinese communist party is reliant on a nationalist narrative to bolster its domestic political legitimacy. A key component of this narrative is standing up for China's rights in the international arena and recovering territories lost to foreign aggression, such as the Senkaku/Diaoyu Islands that China claims to have lost to Japan during the First Sino-Japanese War of 1895. The political symbolic importance of the dispute is raised by the fact that Japan is the primary antagonist of China's nationalist narrative, making it politically costly to appear to be conciliatory or to not to press claims.
- **Security**: The strategic significance of the islands covers a wide range of issues. (1) The islands are important for China's maritime defence strategy to secure control over littoral waters to protect the economically vital but vulnerable coast. A similar strategy can be observed on the South China Sea in China's "Great Wall of Sand". (2) Creating safe patrol areas for nuclear submarines in the absence of direct access to open oceans due to the 1st and 2nd Island Chains, following the Soviet model of safe harbours, is critical to have a high-survivability nuclear second-strike capability. (3) The islands can also be used as forward operating positions during various strategic scenarios, such as a coercive/deterrent campaign against enemy SLOC or a potential campaign against Taiwan.

Based on the high value of the islands, it is not surprising that China has been eager to maintain its claims and has elected to often act in an assertive manner.

China's rational foreign policy choice

If China wants to secure control over the islands, it has four key options to do so:

a) **Use of Force:** China can seek to establish *de facto* control over the islands by going to war with Japan and securing control through military power.
b) **Coercion:** China can attempt to persuade Japan to give up control over the islands by presenting a military or economic threat where the damage would outweigh Japan's interests attached to controlling the islands.
c) **Non-Binding Resolution Options:** China could negotiate with Japan to find a bargain that is acceptable for Japan to surrender the islands. This effort could be facilitated by bringing in trusted third parties to move the proceedings along, for example, through mediation or conciliation.
d) **Binding Peaceful Resolution Options:** China could submit the case to an international tribunal to have the matter settled. In domestic terms this would be similar to suing Japan, where a country would deliver a binding resolution on who owns the islands. The International Court of Justice has a history of settling territorial disputes. The Permanent Court of Arbitration could similarly be used.

Looking back at Part IV, a rational choice approach to decision-making would focus on determining the relative costs and benefits to identify the optimal course of action. This can also be used to determine the possibility of China and Japan going to war. This analysis can be summarized as shown in Table 15.2.

As Table 15.2 shows, none of these options is ideal. Options that could grant the highest benefit, *de jure* control, are virtually unfeasible. The failure is either guaranteed or the risks make it virtually certain. Coercion similarly faces a high risk of failure: Japan has been growing tired of China's assertive politics and being a target of its coercive foreign policy; thus, Tokyo has been eager to stand up to China, a consideration that was an election promise of Prime Minister Shinzo Abe. This leaves one option: If China wants to control the islands, it can only do so reliably through going to war with Japan.

Obviously, war is not an optimal course of action. Its risks are still rather unfavourable. Thus, the rational conclusion for China is to adopt a policy of delaying. It could use the time to improve the feasibility of various courses of action. This conclusion has been originally reached in 2016 and as anticipated China has been following a course of delaying action. In the meantime, the focus on Chinese policy seems to have been twofold:

a) China has been actively building up its navy to achieve a more favourable balance with other regional actors. This is not done solely for the Senkaku/Diaoyu Islands, but in the context of China's general rise as a great power. But altering the balance of power will improve the feasibility

Table 15.2 Foreign policy options

	Cost	*Benefit*	*Risk*
Use of Force	High material costs Reputational penalty for norm following, can affect cooperation and participation in international system	*De facto* control Reputation for strength, can benefit coercion and deterrence	Military balance relatively equal, victory is not guaranteed US participation unknown, incentives for both intervention and non-intervention
	Highest	**High**	**Moderate**
Coercion	Limited material costs (potential demonstrative action) Reputational penalty for norm following, can affect cooperation and participation in international system	*De facto* control	Japan is hardened against coercion Threat without follow-up weakens credibility
	High	**High**	**High**
Non-Binding Peaceful	Domestic political backlash for engaging with Japan	*De jure* control	Japan does not recognize the dispute, so the chance of success is roughly zero No mutually acceptable third party Significant backlash if compromise is unfavourable
	Low	**Highest**	**Not Feasible**
Binding Peaceful	Creates a precedent for other disputes Increases reputational penalty in other disputes if not consistently followed	*De jure* control	China's claims are weak. China cannot demonstrate effective sovereignty in 19th Century. Unlikely a court would overturn a century of continuous possession.
	Low	**Highest**	**High**

of potential use of force over the Senkaku/Diaoyu Islands, increasing the chances of a victory.

b) China has continued its efforts to undermine Japan's case for effective sovereign control. On the one hand, this is important to keep China's claim alive. If a claim is not pursued actively, it will eventually lapse, losing all international validity. While China is deferring a resolution to the dispute

> to a later date, it still needs to maintain its claims by signalling Japan and the rest of the world that it doesn't consider the matter settled. On the other hand, it lays the foundations for China to present a stronger case on the international level of why Japan's policy of not recognizing the dispute is unacceptable. This can be beneficial for potential negotiations or arbitration. Right now, Japan's position is relatively secure because of effective sovereign control. If China can undermine that, then the position of Japan would be significantly weakened.

Delaying offers a lot of strategic utility to China. However, the fundamental fact remains that if it wishes to resolve the dispute, the use of military force still remains the most reliable way to do so. While it is possible that China decides to delay a resolution indefinitely or that political changes in the country would alter these calculations, if the current patterns consist, the dispute's potential for war remains a concern.

Altering calculations

China's adoption of a delaying strategy offers a good opportunity for other stakeholders to engage in preventative measures. The goal would be to divert China from a path where a military resolution to the dispute would be viewed as an optimal outcome. To do this, other stakeholders, such as Japan or the United States, need to examine the rational calculations that would lead China to such a decision. The main focus is to identify key points within these calculations where external pressure can be applied to alter them.

One good example would be deterrence. As discussed above, China is actively working on altering the balance of power with Japan, which would increase the feasibility of a military confrontation, making it a more attractive action. One needs to find ways to counter this. The most straightforward way of doing this is an equal military build-up. By keeping pace with China's growing naval power, Japan can attempt to maintain the current balance of power, preventing China from achieving any favourable gains. The Japanese government under Shinzo Abe has been engaging in such a dynamic, investing significantly into Japan's naval capabilities. The United States also tried to affect China's perceptions of risk: President Obama broke tradition by clarifying the purposefully ambiguous US stance on the islands to state that they fall under the Japanese-American security cooperation. The goal was here to increase China's perceptions that the United States will intervene in a conflict, which would significantly add to the risks of a military campaign. Another avenue for deterrence is to find ways to increase the costs of the action. If China is convinced that devastating international sanctions would be instituted in case of aggression, then it would view the option more negatively. Unfortunately, the international community does not have the best track record with that. Both Russia and North Korea have been defying sanctions and China is counting on the fact that such sanctions against the

People's Republic of China (PRC) would be economically damaging to the rest of the world as well.

An alternative to deterrence, which would focus on disincentivizing unfavourable options through raising costs and risks, would be to seek ways to increase the feasibility of favourable policy choices. As discussed above, the key obstacle to the peaceful resolution of the dispute is that from a Chinese perspective none of those options offer a reliable way to get what they want. While it would weaken Japan's bargaining position, an official recognition for the existence of a dispute could go a long way to reassure China and to reduce the perception that a military resolution is the only possible way. If one utilizes Fravel's model from Chapter 10, escalation is often motivated by a weak bargaining position. Offering a concession to China could offer some stabilization to the situation.

However, ultimately one needs to face the fact that sovereignty is indivisible. If Japan is determined to hold onto the islands, and also seeks to avoid war, then it will need to find ways to address the underlying reasons why China desires the islands in the first place. This would be supremely difficult as China has strong strategic reasons to seek control over littoral waters. The situation is not helped by that the United States under President Obama has engaged in a strategy designed to keep China insecure in order to curb its international ambitions. An insecure country will be increasingly desperate to escape its vulnerable position. Ultimately, for peace to be preserved one needs to recognize that the international order has changed. New powers have arisen whose needs have to be accommodated by the international order. Alternatively, if existing great powers seek to preserve the status quo, then they need to prepare to fight off those who seek to challenge it because of their dissatisfaction with it. Preventative diplomacy in such a scenario would be a delaying strategy in itself: Deterring challengers in the hopes that they either learn to live with the status quo or domestic political changes curb their ambitions.

"A" peace versus "our" peace

It is not difficult to make peace. It is only difficult to make a peace we are willing to live with. If one is simply concerned about the absence of hostilities, then appeasement remains a reliable way to avoid having to go to war. It is a fundamental mistake, often made, to think that the main goal of the international system is to be at peace. War is terrible. But sometimes the alternative is inconceivable. The primary purpose of foreign policy is to realize a state of affairs in the international system that is acceptable to us. Freedom from threat of violence is certainly part of this ideal state of affairs. However, it is not the only component. A broader sense of human security and ideals of human rights are also a significant consideration for the international system. World War II was a choice. Allied power could have chosen to follow a policy of appeasement, giving into the demands of Hitler. Countries like the United States could have stood by and watch Hitler carve out a German *lebensraum* and carry on with broad ethnic cleansing in order to preserve peace. Instead, the Allied forces decided to meet the Wehrmacht on the battlefield to determine whose vision for the international system will be implemented because the values and ideas represented by Nazi Germany were determined to be unbearable, if for nothing else because they threatened a fundamental violation of national sovereignty. Similarly, Nazi Germany or Imperial Japan chose to meet the Allies on the battlefield as they considered the status quo to be fundamentally unacceptable and sought to overturn the system through force. One might argue that this is an extreme example. Not all conflicts are grand struggles for the very soul of the international system, one could argue. But at the end of the day, the same principles apply. All wars are about realizing a specific post-war international order. The purpose of war is not war itself, but the realization of a preferred peace. Even in a limited war context the goal is to realize a specific political objective. The only question is whether one believes strongly enough in their vision of post-war order to be willing to suffer the tragedy of war.

At the same time, peace is not self-sustaining. Once a favourable peace is achieved, it requires constant safeguarding, including going to war to dispatch threats. While humanity may continue to strive to devise international orders that are more equitable, there will always be states that are disenfranchised by the status quo. Even in a system that is simply based on a principle of non-interference, there will always be groups and states that view that it is their role to impose their will on others, based perhaps on religious or moral convictions. In such an instance, if one is invested in preserving the status quo, there might not be another option than to meet the challenger on the battlefield and undertake the sacrifice of war in order to prevent a greater evil.

Peace is based on force, even if that force is held in reserve. While some might respect the international order and the norms of peaceful resolution, others need to be deterred from seeking a forceful overturn of the system through superior military might. Economic coercion, bilateral negotiations, or arbitration are possible because of the understanding that it is not possible for the opponent to simply take what they want without meeting the tragedy of war. Even authors like Mueller recognize that one of the reasons for the supposed peaceful evolution of the international system is the impending apocalypse imposed by the presence of nuclear weapons.

Peace is often viewed as an absence of war. However, this book would make the argument that such an interpretation is incorrect. Peace is war held in reserve. The two concepts are inherently and inseparably interlinked. And this is a key reason why war is never going away or why humanity will never overcome its continued occurrence in the international system. It is inherent to the system. And if war is inherent, then it is absolutely critical to understand it in detail and devise policy that best accounts for it. This book has sought to do that, as it holds the belief that, potentially paradoxically, the route to peace leads through the study of war.

Bibliography

Anderson, B. (2006) *Imagined Communities: Reflections on the Origins of Nationalism.* Verso: London.

Angell, N. (1933) *The Great Illusion*. G.P. Putman's Sons: New York.

Barnett, T. P. M. (2004) *The Pentagon's New Map*. Penguin: New York.

BBC (2019) *Russia deploys Avangard hypersonic missile system.* www.bbc.com/news/world-europe-50927648 (Accessed 14 June 2020).

Biao, Lin, ed. (1966) *Quotations from Chairman Mao Tse-Tung.* 1st edn. Foreign Languages Press: Beijing.

Blackwill, R. D. and Harris, J. M. (2016) *War by Other Means: Geoeconomics and Statecraft*. Harvard University Press: Cambridge.

Braumoeller, B. F. (2019) *Only the Dead: The Persistence of War in the Modern Age.* Oxford University Press: Oxford.

Callahan, William A. (2010) *China: The Pessoptimist Nation.* Oxford University Press: Oxford.

Callahan, W. A. (2012) "Sino-Speak: Chinese Exceptionalism and the Politics of History". *The Journal of Asian Studies* 71(1), pp. 33–55.

Charter of the United Nations. www.un.org/en/sections/un-charter/chapter-i/index.html (Accessed 08 June 2020).

Chase, M. S., Engstrom, J., Cheung, T. M., Gunness, K. A., Harold, S. W., Puska S., and Berkowitz, S. K. (2015) "China's Incomplete Military Transformation: Assessing the Weaknesses of the People's Liberation Army (PLA)". *RAND National Security Division.* www.rand.org/content/dam/rand/pubs/research_reports/RR800/RR893/RAND_RR893.pdf (Accessed 11 April 2016).

Chung, C. (2009) "The 'Good Neighbour Policy' in the Context of China's Foreign Relations". *China: An International Journal* 7(1), pp. 107–123.

Convention for the Amelioration of the Condition of the Wounded in Armies in the Field. Geneva, 22 August 1864. https://ihl-databases.icrc.org/ihl/INTRO/120 (Accessed 08 June 2020).

Convention (IV) respecting the Laws and Customs of War on Land and its annex: Regulations concerning the Laws and Customs of War on Land. The Hague, 18 October 1907.https://ihl-databases.icrc.org/ihl/INTRO/195

Corbett, J. (1911) *Some Principles of Maritime Strategy*. Guttenberg Online.

Convention (II) with Respect to the Laws and Customs of War on Land and Its Annex: Regulations Concerning the Laws and Customs of War on Land. The Hague, 29 July 1899. https://ihl-databases.icrc.org/ihl/INTRO/150 (Accessed 08 June 2020).

Clausewitz, C. (1874) *On War* (Translation by J.J. Graham). Guttenberg.

Cole, B. D. (2010) *The Great Wall at Sea: China's Navy in the Twenty-First Century*, 2nd edn. Naval Institute Press: Annapolis.

De Mesquita, B. B. and Smith, A. (2011) *The Dictator's Handbook*. Public Affairs: New York.

Dupuy, Trevor N. (1979) *Numbers, Predictions and War: Using History to Evaluate Combat Factors and Predict the Outcome of Battles*. Macdonald and Jane's: London.

Fenby, J. (2009) *The Penguin History of Modern China*. Penguin: London.

Fettweis, C. J. (2006) "A Revolution in International Relations Theory: Or, What if Mueller Was Right?". *International Studies Review* 8(4), pp. 677–697.

Fravel, M. T. (2007) "Power Shifts and Escalation: Explaining China's Use of Force in Territorial Disputes". *International Security* 32(3), pp. 44–83.

Goh, E. (2007) "Great Powers and Hierarchical Order in Southeast Asia: Analyzing Regional Security Strategies". *International Security* 32(3), pp. 113–157.

Goodwin, P. H. B. (1999) "The PLA Faces the Twenty-First Century: Reflections on Technology, Doctrine, Strategy, Operations" in J. Lilley and D. Shambaugh (eds.), *China's Military Faces the Future*. M.E. Sharpe: Armonk

Gray, C. S. (2009) *Fighting Talk: Forty Maxims on War, Peace and Strategy*. Potomac Books: Dulles.

Groshkov, S. G. (1979) *The Sea Power of the State*. Pergamon Press: New York.

Handel, M. I. (2001) *Masters of War*, 3rd edn. Frank Cass: London.

Henshall, K. (2012) *A History of Japan: From Stone Age to Superpower*. Palgrave Macmillan: New York.

Hobbes, T. (1651) *Leviathan*. Pandora's Box Classics.

Holmes, J. R. and Toshi Yoshihara. (2008) *Chinese Naval Strategy in the 21st Century: A Turn to Mahan*. Routledge. London, New York.

Huntington, S. P. (1996) *The Clash of Civilizations*. Simon & Schuster: New York.

"Joint Operations Planning" by US Joint Chiefs of Staff (2011). www.dtic.mil/doctrine/new_pubs/jp5_0.pdf (Accessed11 April 2016).

Kaplan, R. D. (2012). *The Revenge of Geography*. Random House: New York.

Kissinger, H. (2012) *On China*. Penguin: New York.

Lilley, J. and Shambaugh, D. eds. (1999) *China's Military Faces the Future*. M.E. Sharpe: Armonk.

Lockheed Martin. (n.d.) *THAAD: Integrated Air and Missile Defense*. www.lockheedmartin.com/en-us/products/thaad.html (Accessed 14 June 2020).

Mearsheimer, J. J. (2014) *Tragedy of Great Power Politics*. Norton: New York.

Ministry of National Defence (PRC) (2015) *Defence Policy - III. Strategic Guideline of Active Defense*. Ministry of National Defence: Beijing. http://eng.mod.gov.cn/Database/WhitePapers/2015-05/26/content_4586711.htm (Accessed 11 April 2016).

Morgenthau, H. (2005) *Politics among Nations: The Struggle for Power and Peace*. McGraw-Hill: London.

Mueller, J. (1990) *Retreat from Doomsday: The Obsolescence of Major War*. Basic Books: New York.

Ng, K. P. (2005) *Interpreting China's Military Power: Doctrine Makes Readiness*. Frank Cass: New York.

Nye, J. S. (2011) *The Future of Power*. Public Affairs: New York.

Pinker, S. (2011) *The Better Angels of Our Nature: A History of Violence and Humanity*. Penguin: London.

Rasler, K. A. and Thompson, W. R. (2006) "Contested Territory, Strategic Rivalries, and Conflict Escalation". *International Studies Quarterly* 50(1), pp. 145–167.

Rawlinson, J. L. (1967) *China's Struggle for Naval Development, 1839-1895*. Harvard University Press: Cambridge.

Russet, B. (1994) *Grasping Democratic Peace*. Princeton University Press: New Jersey, NJ.

Sawyer. R. D. (2007) *The Seven Military Classics of China*. Basic Books: New York.

Schelling, T. (2008) *Arms and Influence*. Yale University Press: New Haven, CT.

Slaughter, A. (2004) *A New World Order*. Princeton University Press: Princeton, NJ.

Swanson, B. (1982) *Eighth Voyage of the Dragon: A History of China's Quest for Seapower*. Naval Institute Press: Annapolis.

Tangredi, S. J. (2013) *Anti-Access Warfare: Countering A2/Ad Strategies.* Naval Institute Press: Annapolis.

The Constitution of Japan (1946) https://japan.kantei.go.jp/constitution_and_government_of_japan/constitution_e.html

The Covenant of the League of Nations (1924) https://avalon.law.yale.edu/20th_century/leagcov.asp (Accessed 08 June 2020).

Till, G. (2004) *Seapower: A Guide for the Twenty-First Century*. Frank Cass: London.

Toshi Y. (2016) "The 1974 Paracels Sea Battle". *Naval War College Review* 69(2), pp. 41–65.

United Nations. *Responsibility to Protect*. www.un.org/en/genocideprevention/about-responsibility-to-protect.shtml (Accessed 08 June 2020).

Wallerstein, I. (2006) *World System Analysis: An Introduction*. Duke University Press: Durham.

Waltz, K. (2010) *Theory of International Politics*. Waveland Press: Long Grove.

Walzer, M. (1977) *Just and Unjust Wars*. Basic Books: New York.

Wang, Z. (2012) *Never Forget National Humiliation.* Colombia University Press: New York.

Wendt, A. (1992) "Anarchy Is What States Make of It: The Social Construction of Power Politics". *International Organization* 46(2), pp. 391–425.

Wiegand, K. E. (2011) "Militarized Territorial Disputes: States' Attempts to Transfer Reputation for Resolve". *Journal of Peace Research* 48(1), pp. 101–113.

Wong, E. (2011, March 30). *China Hedges Over Whether South China Sea Is a "Core Interest" Worth War*. Retrieved 18 May 2011, from the *New York Times*: www.nytimes.com/2011/03/31/world/asia/31beijing.html?_r=4&ref=global-home.

Index

www.ingramcontent.com/pod-product-compliance
Lightning Source LLC
LaVergne TN
LVHW021122300926
849160LV00010B/821

9780367748029